MICROSCOPY HANDBOOKS 47

Electron Diffraction in the Transmission Electron Microscope

Royal Microscopical Society MICROSCOPY HANDBOOKS

Series Advisors

Angela Kohler (Life Sciences), *Alfred Wegener Institut, Notke-Strasse 85, 22607 Hamburg, Germany*

Mark Rainforth (Materials Sciences), *Department of Engineering Materials, University of Sheffield, Sheffield S1 3JD, UK*

Electron Diffraction in the Transmission Electron Microscope

P. E. Champness

Department of Earth Sciences, University of Manchester, Manchester M13 9PL, UK

Taylor & Francis
Taylor & Francis Group

LONDON AND NEW YORK

© Taylor & Francis 2001

First published by Taylor & Francis
2 Park Square, Milton Park, Abingdon, Oxon, OX14 4RN
270 Madison Ave, New York NY 10016

Transferred to Digital Printing 2008

A CIP catalogue record for this book is available from the British Library.

ISBN 1 85996 147 9

Production Editor: Paul Barlass.
Typeset by Marksbury Multimedia Ltd, Midsomer Norton, Bath, UK.

Front cover: [001] Tanaka CBED pattern of the 24R polysome of the Al,
Sn oxide mineral nigerite (see *Figure 6.1*) and [111] schematic Kikuchi
map for diamond cubic crystals (see *Figure 4.7*).

Publisher's Note
The publisher has gone to great lengths to ensure the quality of
this reprint but points out that some imperfections in the
original may be apparent.

Contents

7. The fine structure in electron diffraction patterns 113

Appendices 133

Index 167

Abbreviations

APB	antiphase domain boundary
CBED	convergent-beam electron diffraction
CCD	charge-coupled detector
FEG	field-emission gun
FOLZ	first-order Laue zone
FWHM	full width at half maximum
FWTM	full width at 10th maximum
HOLZ	higher-order Laue zone
K-M	Kossel–Möllenstedt
relp	reciprocal lattice point
relrod	reciprocal lattice rod
SAD	selected-area diffraction
SOLZ	second-order Laue zone
STEM	scanning transmission electron microscope
TEM	transmission electron microscope
ZAP	zone-axis pattern
ZOLZ	zero-order Laue zone

List of symbols

a, b, c	lattice parameters
a^*, b^*, c^*	reciprocal lattice parameters
A	amplitude
C_s	spherical aberration coefficient
d	spacing of periodic object
d_{hkl}	spacing of lattice planes (hkl)
D	diameter of diffraction aperture
E	energy
f	atomic scattering factor
f	focal length of lens
F_{hkl}	structure factor
$\mathbf{g}_{hkl}$	reciprocal lattice vector
G	radius of HOLZ rings
hkl	indices of reflection from (hkl) planes
(hkl)	Miller indices of a plane
$\{hkl\}$	form or set of planes related by symmetry
H^*	spacing of reciprocal layers
I	intensity
$\mathbf{k_0}$	vector of the incident beam
$\mathbf{k}$	vector of the diffracted wave
L	camera length
M	magnification
m	pitch of screw axis
n	integer
N	$h^2+k^2+l^2$
N	total number of atoms in unit cell
N	order of screw axis
nm	nanometre (10^{-9} metres)
pm	picometre (10^{-12} metres)
R	distance in the diffraction pattern as measured on the film or screen
$\mathbf{s}$	deviation parameter
s'	effective deviation parameter
t	thickness of specimen
t	lattice repeat
u	fourth crystallographic axis used in hexagonal and trigonal systems

$[UVW]$	zone direction
$<UVW>$	family of zones related by symmetry
v	velocity
V	voltage
V_c	volume of unit cell
x	distance
y	displacement
x, y, z	direction of crystallographic axes
x_n, y_n, z_n	fractional coordinates of nth atom in unit cell
x*, y*, z*	direction of reciprocal crystallographic axes

Greek symbols

α	primary tilt angle
α	scattering angle
2α	convergence angle
α, β, γ	angles between crystallographic axes
$\alpha^*, \beta^*, \gamma^*$	angles between reciprocal crystallographic axes
β	secondary tilt angle
Δ	path difference
Δf	defocus
ΔR	displacement of Kikuchi line from diffraction spot on film or screen
ϕ	angle between **g** vectors/normal to planes
ϕ	phase difference
ϕ_{hkl}	phase angle for reflection hkl
ξ	extinction distance
λ	wavelength
Ψ_{hkl}	amplitude of diffracted beam
Ψ_0	amplitude of undeviated beam
θ	Bragg angle

Preface

This text is the result of many years' experience of teaching the elements of electron diffraction to postgraduate students in the University of Manchester and UMIST, and to students attending the annual Royal Microscopical Society School in Electron Microscopy. I have tried to minimize the use of mathematics, but where I considered mathematics unavoidable, I have made the derivations as full as possible. I have also tried to illustrate physical concepts with suitable diagrams, wherever appropriate.

My examples have been taken from a range of metals, minerals and ceramics, and I have tried to make the coverage as broad as possible in terms of crystal symmetry. My experience has been that students who are only familiar with the cubic and hexagonal systems come unstuck when confronted with a system of lower symmetry! I also hope that biologists will find the book accessible.

The appendices cover material that is not dealt with in the main text. They include a summary of essential crystallography and an introduction to space groups. Other appendices and parts of the text provide tables of data and crystallographic formulae that I hope will provide the reader with a useful source of reference.

P.E. Champness

Acknowledgements

I am extremely grateful to the following people who have made helpful comments on various parts of the text: Alwyn Eades, Chris Hammond, Peter Kenway, Gordon Lorimer and Jian-Guo Zheng. I am also grateful to Sue Maher, Robin Hadley and Margaret Banton, who printed the photographs, and to Richard Hartley who produced the line drawings from my, often rough, originals. Feridoon Azough, Chris Hammond, Peter Kenway, Gordon Lorimer and Phil Prangnell kindly supplied photographs.

A number of copyright holders have permitted me to reproduce figures, the acknowledgement of which is made in the figure captions.

Safety

Attention to safety aspects is an integral part of all laboratory procedures and both the Health and Safety at Work Act and the COSHH regulations impose legal requirements on those persons planning or carrying out such procedures.

In this and other Handbooks every effort has been made to ensure that the recipes, formulae and practical procedures are accurate and safe. However, it remains the responsibility of the reader to ensure that the procedures which are followed are carried out in a safe manner and that all necessary COSHH requirements have been looked-up and implemented. Any specific safety instructions relating to items of laboratory equipment must also be followed.

1 Diffraction and the electron microscope

When we look at a distant street light through a finely woven net curtain or an umbrella we see a pattern of bright spots. This pattern is not an image, it is a diffraction pattern. We see a regular array of spots because the fabric of the curtain or the umbrella also has a regular, periodic weave. The pattern that we see is telling us about the periodicity of the weave and its orientation; if you rotate the fabric about a direction perpendicular to its surface, the pattern of spots will also rotate. We will look at this phenomenon in more detail at the end of the next section.

1.1 How a lens forms a diffraction pattern

We are most familiar with the use of lenses to form a (usually) magnified image of an object; for instance, with a magnifying glass or an optical microscope. However, any converging lens used in this way is also forming a diffraction pattern. What happens can be understood by reference to *Figure 1.1*. Here we have a periodic object (it could be a piece of net curtaining) placed in front of a converging lens which is illuminated by a parallel beam of monochromatic (single wavelength) light. Some of the light is transmitted undeviated through the object and some is 'scattered' (some will also be absorbed). The undeviated light travels parallel to the optic axis of the lens and is focused by the lens so that it passes through the **principal focal point** F_0 at a distance from the lens equal to the focal length, f. A plane drawn through F_0 perpendicular to the axis is the **back focal-plane** of the lens (note that there is a **conjugate**, i.e. optically equivalent, **front focal-plane** at a distance f from the lens on the object side of the lens). Light scattered at angles $\pm \alpha$ by the object will form parallel beams which are focused to points F_1 and F_1' in the back focal-plane, at a distance from the axis proportional to the angle α. This pattern of spots formed when the transmitted and scattered beams are focused into the back focal-plane is the diffraction pattern. In fact, for an infinite, periodic object such as our net curtain the spots are only formed for rays that are scattered in specific directions (we will learn

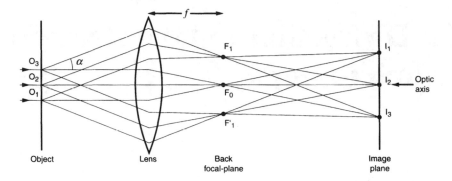

Figure 1.1. Formation of an image and diffraction pattern by a lens from an infinite periodic object. For parallel, monochromatic illumination, the transmitted and diffracted beams are focused into spots, F_0, F_1, F_1' etc., which form the diffraction pattern in the back focal-plane of the lens at a distance f, the focal length, from the lens. Only three diffraction spots are shown for clarity. The diffracted beams recombine to form a magnified image in the image plane. Note that the image is inverted with respect to the object, but the diffraction pattern is not. In effect, the screen of the electron microscope is in the plane I_1–I_3 in the imaging mode and in the plane F_1–F_1' in the diffraction mode.

the rule that governs these directions later). The rays that form the diffraction spots in the back focal-plane go on to form an inverted image in the image plane. If you follow the three rays shown in *Figure 1.1* that are emerging from any of the three apertures O_1, O_2 or O_3 of the object, you will find that they end up at the same point in the image (rays from O_1 end up at I_1 for example, as you would expect), each having passed through a *different* diffraction spot. In other words, information about the object is contained in *each* diffraction spot. A direct consequence of this effect is that, in order for the lens to form an image of the object (i.e. in order to resolve the weave of the curtain material), at least two diffracted beams should enter the objective lens and be allowed to recombine in the image plane. This principle was first described by Ernst Abbe in the 19th century and is known as the **Abbe criterion**.

If we explore a little further, we find that there is a definite relationship between the periodicity and orientation of the object and the spacing and orientation of the spots in a diffraction pattern. You can easily see what this relationship is if you hold a set of meshes, net-curtain materials or diffraction gratings of varying periodicity in front of a laser pointer (do NOT look directly into the laser beam) or other small torch and view the diffraction pattern produced on the wall on the opposite side of the room. You will find that the spacing of the diffraction spots is inversely proportional to the object's periodicity (for instance, if you halve the spacing of the grating or mesh, the spacing of the diffraction spots doubles) and that the separation of the spots produced by the grating or by the weave of the net is perpendicular to the lines of the grating (*Figure 1.2*). (If you do not have a set of such objects, you can see the size effect by rotating a mesh about the direction of one of the threads, thus effectively shortening the mesh periodicity perpendicular to the direction of rotation.

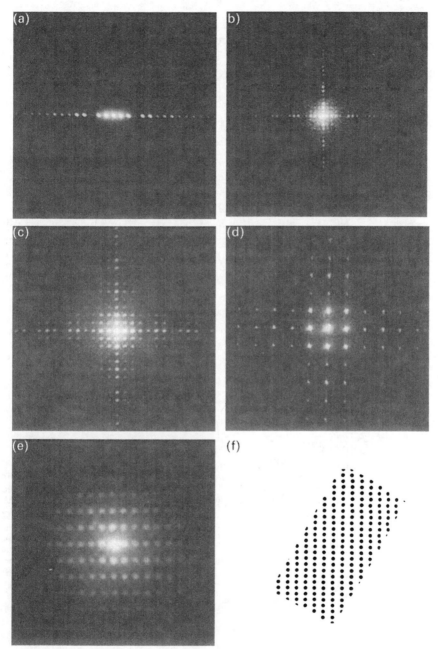

Figure 1.2. Optical diffraction patterns taken with a laser and recorded under the same conditions. (a) Pattern from a diffraction grating with spacing 0.126 mm. The lines of the grating were parallel to the length of the page; notice that the row of diffraction spots is perpendicular to this direction. (b) Pattern from a 100-mesh electron-microscope grid, grid spacing 0.25 mm. (c) Pattern from a 200-mesh electron-microscope grid, grid spacing 0.125 mm. (d) Pattern from a 400-mesh electron-microscope grid, grid spacing 0.0625 mm. (e) Diffraction pattern from the rectangular two-dimensional 'crystal' shown in (f). Note the reciprocal relationship between the spacings in the objects and the spacing of the corresponding row of diffraction spots. Note also that the row of spots is perpendicular to the grating from which it is diffracted.

The spacing of the corresponding diffraction spots will increase). What we have been looking at in the diffraction pattern is also known as the **reciprocal lattice** because of the inverse (reciprocal) relationship it bears to the **direct lattice** – the object.

1.2 Introduction to diffraction in the TEM

A transmission electron microscope (TEM) consists of a source of electrons (the electron gun) and a series of electromagnetic lenses, as shown in *Figure 1.3* (see also *Table 1.1*). The most critical components of a magnetic lens are the soft-iron pole-pieces which produce an axially symmetric magnetic field for focusing the electrons. The rest of the lens is a magnetic yoke containing the windings for energizing the lens with a d.c. current. By varying this current, the magnetic field, and hence the

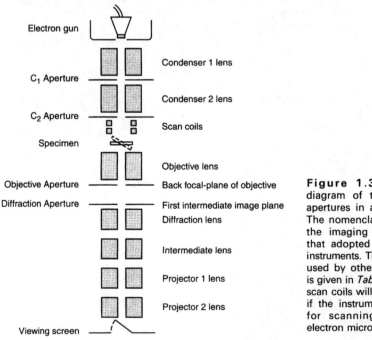

Figure 1.3. Schematic diagram of the lenses and apertures in a modern TEM. The nomenclature shown for the imaging lenses follows that adopted for FEI/Philips instruments. The nomenclature used by other manufacturers is given in *Table 1.1*. Note: the scan coils will only be present if the instrument is designed for scanning transmission electron microscopy.

Table 1.1. Nomenclature used by different manufacturers for the imaging lenses in a TEM

FEI/Philips	JEOL	Hitachi
Objective	Objective	Objective
Diffraction	First intermediate	First intermediate
Intermediate	Second intermediate	Second intermediate
Projector 1	Third intermediate	First projector
Projector 2	Projector	Second projector

focal length of the lens, is changed. This ability to change the focal length is one of the most important ways in which magnetic lenses differ from glass ones. Another difference is that, unlike the glass lenses in a light microscope, magnetic lenses in an electron microscope, because of their construction, size and weight, cannot be moved physically with respect to one another in order to focus the image, although the specimen can be moved up or down in the holder to achieve a certain amount of focusing. We will see some other differences between the two types of microscope later.

The specimen is placed in a holder in the front focal-plane of the objective lens (*Figure 1.3*), i.e. at a distance f in front of the lens. *Figure 1.4* shows schematically the ray paths for the imaging system of a TEM that employs three stages of magnification. The magnified image produced by the objective in its image plane serves as an object for the diffraction lens, which produces a second intermediate image, which is further magnified by the projector lens to produce the final image on the fluorescent viewing screen (or on the photographic film which is just

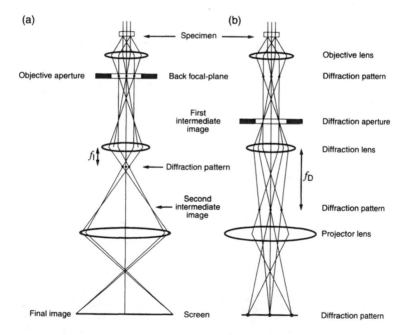

Figure 1.4. Schematic diagram showing the imaging system of a TEM. A three-stage magnification system is shown. (a) The ray paths for the imaging mode in which the image formed by the objective lens forms the object for the diffraction lens. Note: in this mode there is an objective aperture in the back focal-plane of the objective lens which restricts the number of diffracted beams that recombine in the image plane. This arrangement improves contrast in the image. (b) The ray paths for the diffraction mode. Here the objective aperture is withdrawn and a diffraction aperture is inserted in the image plane of the objective lens (the first intermediate image). The diffraction lens is weakened (its focal length is changed from f_1 to f_D) so that its object plane is the back focal-plane of the objective lens. Hence, the diffraction pattern is magnified by the projector lens and the diffraction pattern appears on the screen of the instrument. Note that the top part of the diagrams (from the specimen to the first image plane) is identical to *Figure 2.1*.

below the screen). This is the imaging mode (*Figure 1.4a*); the action of the objective lens and the diffraction/projector lenses parallels that of the objective and eyepiece, respectively, in the optical microscope. However, the diffraction lens can be reduced in strength (its focal length is increased from f_I to f_D) by decreasing the current in the windings so that an image of the diffraction pattern in the back focal-plane of the objective lens is focused on the final screen (*Figure 1.4b*); in other words the 'object' for the diffraction lens in the diffraction mode is the diffraction pattern produced by the objective lens. Referring back to the diagram of a simple lens in *Figure 1.1*, the screen of the electron microscope is effectively in the plane $I_1–I_3$ in the imaging mode and in the plane $F_1–F_1'$ in the diffraction mode. The weakening of the current in the diffraction lens is simply achieved by selecting the diffraction mode on the microscope's console. If an aperture of diameter D (called the selected-area or diffraction aperture) is placed in the first image plane (*Figure 1.4b*), and if the objective lens behaves as a 'perfect' lens (i.e. one without aberrations; see Section 1.5), only those electrons passing through an area of diameter D/M in the specimen will reach the final screen, where M is the magnification of the objective lens. This can be seen in *Figure 1.5* which shows the **virtual aperture** produced in the object plane. In practice, D may be about 50 μm and since M is about 50, the diameter of the area selected is about 1 μm. The diffraction pattern from this area only is therefore observed. This technique is known as **selected-area**

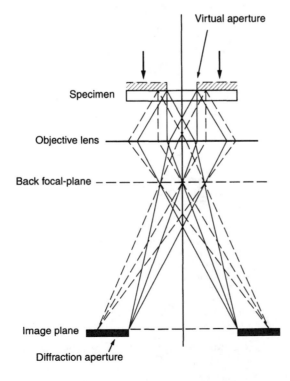

Figure 1.5. Ray diagram showing the formation of a selected-area (SAD) pattern. The insertion of a (diffraction) aperture in the image plane of the objective lens results in the creation of a 'virtual' aperture in the object plane. Only electrons falling inside the virtual aperture will pass through into the imaging system; all other electrons (for instance the dashed rays) will hit the diffraction aperture and will not contribute to the image or the diffraction pattern seen on the screen. If the diameter of the diffraction aperture is D and the magnification of the lens is M, then the area selected in the object has diameter D/M. Diagram courtesy of P.B. Kenway.

diffraction or **SAD**. It allows observation of diffraction patterns from small areas of the specimen so that, among other things, a correlation between features seen in the image and the crystallography of the specimen can be made. Unfortunately, the technique is subject to certain errors, both systematic and random, which are described in Section 1.5. These errors, which are in part due to aberrations in the lens, limit the minimum size of the area that can be selected.

Modern TEMs usually have at least two condenser lenses to focus the image of the electron source onto the specimen and five imaging lenses (*Figure 1.3*) which are given varying names by different manufacturers; I will adopt the nomenclature used for FEI/Philips instruments. This larger number of imaging lenses allows a greater range of magnification, both of the image and of the diffraction pattern, than is possible with the three-lens arrangement in *Figure 1.4*.

1.3 SAD in the TEM

Once you have chosen the area from which you wish to obtain a diffraction pattern, you only need to: insert a diffraction aperture, select the diffraction mode, select the required **camera length** (the magnification of the diffraction pattern), remove the objective aperture and focus the pattern. An example of a diffraction pattern from a single crystal taken at two different camera lengths is shown in *Figure 1.6*. If the area selected is to be accurate, it is important that the image of the specimen is accurately focused in the plane of the diffraction aperture (see Section

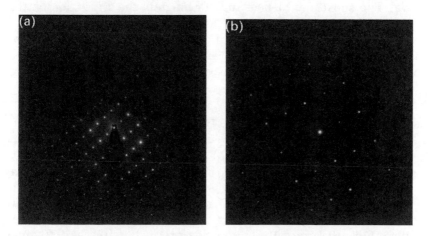

Figure 1.6. Diffraction patterns of the mineral titanite ($\sim CaTiSiO_5$), which is monoclinic ([100] zone axis), taken at camera lengths of (a) 707 nm and (b) 1.43 m, allowing for print size. Note the rotation of about 15° clockwise between the patterns in (a) and (b).

1.6). The diffraction and the projector lenses are programmed within a given range of magnification (the SA, or selected-area, range) so that the plane of the diffraction aperture (the first image plane) is always focused on the viewing screen. At the same time, the objective lens operates at a more-or-less fixed excitation so that it focuses the image in that plane.

Further experimental details of how to obtain a SAD pattern are given in Section 1.9.

1.4 Low- (or small-) angle diffraction

Some biological and other organic crystals have atomic repeats much greater than 10 nm, and the range of camera lengths available by normal SAD may only give a small diffraction pattern from them (remember the reciprocal relationship that we noticed in Section 1.1 between the object and its diffraction pattern; more detail is given in Section 2.3). For instance, a spacing of 100 nm in the object requires a camera length of 54 metres for 100 kV electrons, if the spacing of the spots in the pattern is to be 2 mm (for how this value is arrived at, see Section 2.3), a minimum value for easy measurement on the negative. This camera length is about eight times larger than the maximum normally available in the SAD mode.

Larger camera lengths can be obtained using a different procedure. Here, the objective lens is switched off or reduced to a minimum and the condenser system is focused on the object plane of the diffraction lens. A small **spot size** (the nominal diameter of the beam at the object), obtained by a strongly excited C1 lens, and a small C2 aperture are essential. The subsequent lenses further magnify the diffraction pattern, and camera lengths of the order of several hundreds of metres can be obtained. Details of the procedure differ from one microscope to another, but the technique can be used with any microscope for which there is a flexibility in control over all the lenses.

1.5 Limitations to the accuracy of SAD

The area of the specimen defined by the diffraction aperture in the image plane of the objective lens will not generally correspond exactly to the area from which the associated diffraction pattern comes. There are three sources of error. Firstly, if the image is incorrectly focused, the plane of the diffraction aperture will not coincide precisely with the first image plane and diffracted rays originating from outside the area selected by the aperture will contribute to the pattern. The error involved is proportional to Δf, where Δf is the focusing error. A similar error will occur if the specimen is not in the **eucentric plane** (when the specimen is in this plane and the image is in focus, the objective lens current is an optimum value). This is because the focal length of the objective lens, even when accurately focused on the object, will differ by some value Δf from its correct value if the specimen is at an incorrect height. Adjustment of the specimen height is possible on all modern microscopes.

The third error in the area selected is caused by **spherical aberration** in the objective lens; a phenomenon that can only currently be corrected for with a customized (and expensive) lens (Haider *et al.*, 1998; Krivenek *et al.*, 1999). Spherical aberration is a lens defect that causes electrons scattered furthest from the optic axis to be brought to focus closer to the lens than rays that are near-axial (*Figure 1.7*); in other words, the focal length f is shorter for rays that are scattered at a high angle. Thus, even if you manage to focus the image accurately, diffracted beams coming from areas outside the selected one will contribute to the diffraction pattern. The effect is shown in *Figure 1.8*, which shows one diffracted beam and the undeviated beam. The undeviated rays from the area x–y in the specimen (the dot-dashed lines) are focused through the principal axial focal-point at O in the back focal-plane at a distance equal to the focal length, f_0, from the lens. If the lens were perfect, the rays diffracted by the specimen at an angle α to the optic axis would follow the path shown by the dotted lines, after passing through the lens, and the diffraction spot would appear at B in the back focal-plane. Both the undeviated and diffracted rays would be focused in the normal image plane at YX (note the inversion of the image with respect to the object, as is also apparent in *Figure 1.1*). A diffraction aperture placed at YX in the image plane would then select the area x–y in the specimen.

For a lens with spherical aberration, however, the diffracted ray is more strongly focused along the full line to position C, which is slightly closer to the lens than B and lies in a different (back focal-) plane at a distance f_α from the lens. The result is that the image of the object x–y formed by the rays entering the objective lens at angle α is at Y′–X′; i.e. the end points of the image are displaced distances XX′ and YY′ (which are unequal) in the plane of the diffraction aperture relative to the image

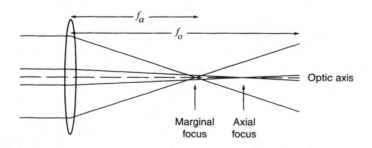

Figure 1.7. Spherical aberration in a lens. Marginal rays are brought to focus nearer to the lens than near-axial rays. The former therefore have a shorter focal length, f_α, than the latter, f_0.

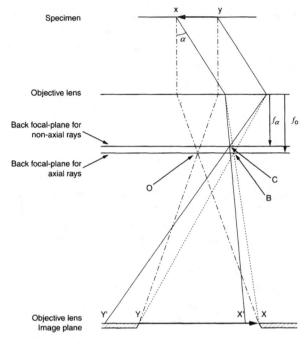

Figure 1.8. The effect of spherical aberration of the objective lens on the accuracy of the area selected by the diffraction aperture. The diagram shows one diffracted beam at an angle α to the optic axis and the undeviated beam. B is the position of the diffraction spot for a perfect lens and C is the position of the spot for a lens with spherical aberration. C is slightly closer to the objective lens than O and B because the focal length, f_α, is slightly shorter for marginal rays than the focal length for axial rays, f_0. The image of the object x–y formed by the diffracted rays is at Y'–X', whereas the image formed by the undeviated beam is at Y–X. The displacement between the two images is proportional to $MC_s\alpha^3$ where M is the magnification of the objective lens and C_s is the spherical aberration constant for the lens. Diagram courtesy of P.B. Kenway.

from the undeviated beam. These distances are proportional to $MC_s\alpha^3$ where M is the magnification of the lens and C_s is the spherical aberration constant of the lens. So, for a lens with spherical aberration, the area selected in the object using the diffraction aperture corresponds to XY *only* for the direct beam; it is displaced by a distance proportional to $C_s\alpha^3$ for each diffracted beam; in other words each diffracted beam 'selects' a different area of the specimen and the error increases very rapidly with scattering angle because of its dependence on the cube of α. Thus there is little point in trying to isolate an area smaller than about 0.5 μm in the specimen. If the magnification of the objective lens is 40 times, this figure translates to an aperture size of 20 μm in the first image plane. Even with a high-resolution TEM (very low C_s) operating at 300–400 kV, the area that can be isolated for SAD is no better than 0.1 μm.

The accuracy of SAD can be improved if the operating voltage is increased. This is a consequence of the fact that the scattering angle α is proportional to λ, the wavelength of the radiation, and λ decreases with voltage (see Appendix D). However, the improvement is not spectacular as λ decreases from 3.7×10^{-3} nm for 100 kV electrons to 1.64×10^{-3} nm for 400 kV electrons, a change of 56%. A more effective (and cheaper!) alternative is described in the next section.

1.6 Diffraction using small probes

We can achieve a dramatic decrease in the size of the area from which the diffraction pattern comes by focusing the electron beam down onto the specimen and dispensing with the diffraction aperture in the first image plane. The beam is no longer parallel and the area selected is, to a first approximation, equal to the size of the beam, known as the **probe**. Convergent-beam electron diffraction (CBED), as the technique is generally known, was first discovered in 1939 by Kossel and Möllenstedt, but has only been developed since the 1970s with the advent of instruments designed for high-resolution microanalysis. The only major problem with using this technique arises in beam-sensitive specimens (see Section 1.8.2) because of the high current densities in the focused electron beam. For such specimens CBED may be impossible, even with the use of a cooling stage and high voltages.

As we have seen, in conventional SAD an approximately parallel electron beam is incident on the specimen, and the resulting diffraction pattern from a periodic sample consists of an array of sharp spots in the back focal-plane of the objective lens (*Figures 1.1, 1.4, 1.9a*). If, instead, the incident beam is convergent (*Figure 1.9b*), the spots become discs and their diameters depend upon the convergence angle 2α, which, in turn, is controlled by the size of the C2 aperture and the lens configuration (see below). *Figure 1.10* shows diffraction patterns in which four different convergence angles were used. As you can see, if the convergence angle is large, the discs overlap and it can be difficult to measure the periodicities in the pattern. In fact, these wider-angle patterns can give us a lot of useful crystallographic information (a topic that will be covered in Chapter 6), but for the moment we will confine ourselves to the situation in which the discs do not overlap and will refer to such patterns as **microdiffraction patterns**. Sometimes you will come across the term **Kossel–Möllenstedt** (or K-M) patterns for such patterns and the term **Kossel patterns** for the situation when the discs overlap.

In modern TEMs, a convergent probe may be formed in most operating modes; i.e. the **TEM mode** (**microprobe** and **nanoprobe**) and the **STEM** (scanning transmission electron microscope) mode. You can form a CBED pattern in the TEM mode in any TEM that can produce a small probe ($\leqslant 1$ μm in size) with a convergence angle of > 20 mrads. This may not be possible on TEMs made before the late 1970s. Most modern TEMs use a condenser-objective lens, as illustrated in *Figure 1.11*. The specimen is placed in the centre of the very strong magnetic field

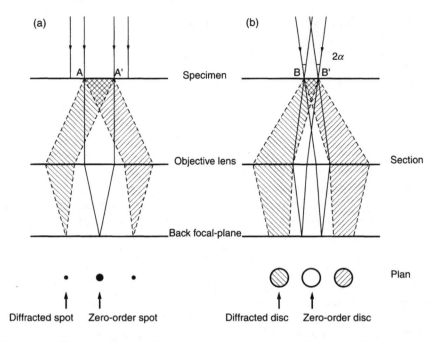

Figure 1.9. The formation of (a) a conventional (selected-area) electron diffraction pattern and (b) a convergent-beam electron diffraction pattern. In (a) the area AA' is selected by means of an aperture in the first intermediate image plane (see *Figure 1.4b*), whereas in (b) the area BB' selected is determined by the probe size. The diameter of the discs in the CBED pattern depends on the convergence angle 2α.

between the pole pieces of the objective lens, the upper part of which, the **prefield lens**, acts as a very strong condenser lens immediately in front of the specimen to focus the electron beam onto the specimen. The second half of the lens is the objective lens that forms the image and diffraction pattern of the specimen. The **auxiliary** lens (or **mini-lens**) above the prefield lens in *Figure 1.11* has the effect of partially nullifying the converging effect of the prefield lens, allowing a larger area to be illuminated in the imaging mode. When a CBED pattern is formed in the TEM mode all the lenses in the illumination system are used (*Figure 1.11a*) and the electron beam is focused on the specimen using the C2 lens.

In the STEM mode of modern TEMs the auxiliary lens is switched off, the first condenser lens is strong (small spot size) and the second condenser lens is either very weak or switched off. In the STEM imaging mode a small probe is scanned over the area of interest using two pairs of scan coils situated below the second condenser lens as shown in *Figure 1.3* (see Keyse *et al.* 1998 for an introduction to STEM techniques). Clearly your TEM must be fitted with a STEM system for you to be able to do this! In this mode the electron beam is focused on the specimen using the objective lens. The convergence angle at the specimen in the STEM mode is several times larger than that obtained in the TEM mode for the same C2 aperture (*Figure 1.11b*).

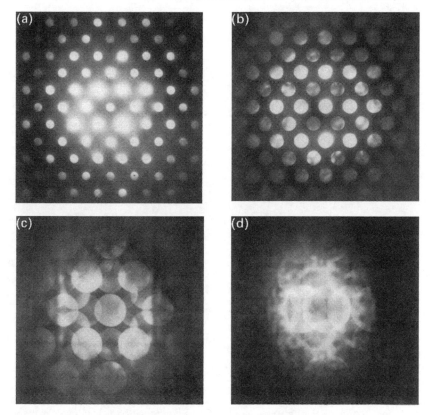

Figure 1.10. Convergent-beam electron diffraction patterns of synthetic Mg_2GeO_4 spinel (F cubic, $a = 0.825$ nm) taken using convergence angles of: (a) 1.4×10^{-3}, (b) 2.3, (c) 4.9 and (d) 8.1 milli-radians. (a) and (b) are referred to as microdiffraction or Kossel–Möllenstedt (K-M) patterns as their diffracted discs do not overlap; they were taken in the 'microprobe mode' of a Philips 400T TEM using 30 and 50 μm C2 apertures, respectively. (c) and (d) were taken in the 'nanoprobe mode' with 30 and 50 μm C2 apertures, respectively. The zone axis was near [110].

In the Philips 400 series of microscopes a so-called high-resolution STEM mode, in which both the C2 and auxiliary lenses are switched off, is also available for CBED. Note that in older machines STEM attachments were available, but the electron optical design varied from machine to machine.

In the FEI/Philips range of microscopes (the 400, CM and Tecnai series), a **nanoprobe** mode is available which uses the same lens configuration as in STEM operation. Here, however, C2 is used to focus the beam onto the specimen, as in the TEM mode. The convergence angles produced in the nanoprobe mode are intermediate in size between those in the TEM and STEM modes for the same size of C2 aperture. The convergence angles produced in the STEM and nanoprobe modes will be too large to produce a microdiffraction pattern in many samples (the patterns shown in *Figure 1.10c* and *d* were both taken in the nanoprobe mode).

Details of how to obtain CBED patterns are given in Section 1.9.

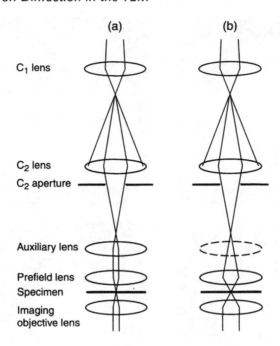

Figure 1.11. Schematic ray diagrams for probe formation in a TEM with a condenser-objective lens. The objective lens consists of two parts: the prefield and the imaging objective lenses; the specimen sits between the two. The auxiliary lens has the effect of partially nullifying the converging effect of the prefield lens, allowing a larger area to be illuminated in the imaging mode. (a) TEM mode, (b) mode in which the auxiliary lens is switched off. In STEM operation, the beam is focused using the objective lens. This lens configuration is also used in the so-called nanoprobe mode on FEI/Philips instruments. In the nanoprobe mode the C2 lens is used to focus the beam. Note that, although the focal lengths of the auxiliary and prefield lenses are shown as roughly equal, in reality the former is several times larger than the latter. Adapted from Champness PE, *Mineral. Mag.*, 1987, with the permission of the Mineralogical Society, London.

1.7 Spatial resolution

What spatial resolution can be achieved using CBED methods? I have said earlier that the resolution is, to the first approximation, equal to the diameter of the electron probe; but what is that? You will find, in manufacturers' brochures and manuals, tables showing nominal probe sizes for different spot sizes and operating conditions (TEM or STEM), but what exactly do these values mean? They are normally the calculated values at 'full width at half maximum', FWHM; in other words, the width of the probe at half its maximum intensity, and assume that the intensity profile of the beam is Gaussian (in a Gaussian probe 50% of the total current is contained within a disc of diameter equal to the FWHM and 90% is contained within the full width at 10th maximum, FWTM). This assumption may be far from the truth. The images of focused electron probes under various conditions are shown in *Figure 1.12*, together with their profiles. The probe on the far left is approximately Gaussian in shape, but the others are not. The 'tails' in the intensity distributions of the other probes have been produced by spherical aberration because the probes are highly converging (remember that spherical aberration causes electrons travelling furthest from the optic axis to be brought to focus closer to the lens than rays that are near-axial). For the probe in *Figure 1.12d* the tail contains an estimated 95% of the total probe current and extends to 1000 nm, despite having a FWHM of only 10 nm. If the probe size is defined as the diameter that contains 90% of the beam current, the spatial resolution for CBED when such a STEM probe is used can be more than 40 times the FWHM, compared with 1.82 times the FWHM for a Gaussian probe.

One way of decreasing the convergence angle is to use a smaller C2 aperture (*Figure 1.11*), at the expense, of course, of beam intensity. But it

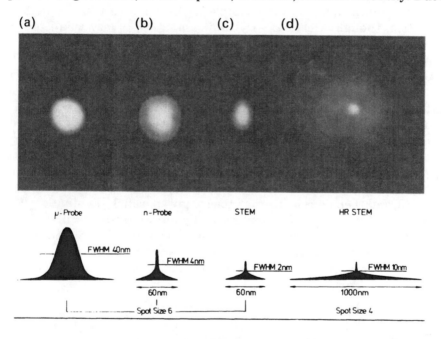

Figure 1.12. Micrographs and schematic drawings of the electron density distributions in electron probes formed under various operating conditions in a Philips EM 400T. A 70-μm C2 aperture was used in each case, and the same excitation of C1 was used in (a), (b) and (c). In (d) a lower excitation of C1 was used. (a) In the 'conventional' TEM, or microprobe, mode the convergence angle is low and the probe is essentially Gaussian. (b) In the TEM nanoprobe mode, when the auxiliary lens is switched off, the convergence angle is greater and there is significant spherical aberration. This results in a non-Gaussian probe. (c) The STEM mode shows similar characteristics to the nanoprobe mode. (d) A STEM probe with an even higher convergence angle gives rise to an even larger 'tail' in the electron distribution. In (b), (c) and (d) the tails would be reduced if a smaller condenser aperture had been used. Reproduced from Cliff G and Kenway PB, *Microbeam Analysis*, 1982, with the permission of the San Francisco Press Inc.

will not be possible to use the ideal size of aperture for each operating condition (unless your microscope has a facility for continually varying α, see Section 1.10.1) as there are usually only three or four apertures to choose from. To minimize the effects of spherical aberration, in addition to using the most appropriate aperture, the beam must be well aligned, the condenser astigmatism corrected and all the apertures in the illumination system accurately centred. If you are concerned with knowing the characteristics of your probe, the best solution is to photograph it and obtain its profile using a microdensitometer on the photographic image, as shown in *Figure 1.12*.

Another factor that affects the size of the area from which a microdiffraction pattern comes is the thickness of the specimen. This is because, as the beam penetrates the specimen, it spreads out. The beam spreading is proportional to the thickness of the specimen to the power $^3/_2$ and inversely proportional to the accelerating voltage (see for instance Williams and Carter, 1996). Thus, it is sensible to use as high a voltage and as thin a sample as possible to attain the highest resolution, though of course very thin samples and very small probes will not produce very intense diffraction patterns and will require long exposures.

Although I have sounded a note of caution about nominal 'probe sizes' and beam broadening, it is still possible to obtain microdiffraction patterns from areas as small as 2 nm with current instrumentation and a field-emission gun (FEG). With a LaB_6 gun the equivalent value is about 10 nm.

In modern TEMs, diffraction patterns with sharp spots can be obtained from areas a few tens of nanometres in diameter by first creating a small probe in TEM or nanoprobe mode, by using a small C2 (30 µm or less) aperture, and then defocusing the probe to give parallel illumination. The **Riecke** diffraction pattern, sometimes called **micro–micro diffraction**, is then seen when the diffraction mode is selected. There are also a number of scanning methods for obtaining diffraction patterns from small areas which are outside the scope of this Handbook, but are covered by Williams and Carter (1996) and Keyse *et al.* (1998).

1.8 Problems with specimens

1.8.1 Specimen preparation

The standard methods for preparing biological specimens for the TEM are covered by Misell and Brown (1987), and the methods for non-biological specimens are covered by Goodhew (1984) and Williams and Carter (1996). Artefacts may arise in the preparation process that introduce spurious detail into the electron diffraction pattern.

Negative staining of biological samples may result in modification of the structure of the sample and the stain itself may recrystallize under the electron beam. One solution is to dispense with the stain and use a supporting medium such as glucose; an even better method is to observe the specimen in the frozen-hydrated state or to use an environmental cell in which the specimen may be observed in the wet state (Misell and Brown, 1987).

Electropolishing of metals sometimes produces a crystalline and/or amorphous oxide film on the surface of the specimen, for example on Mg and Al, and ion thinning may also re-deposit material on the surface of non-metallic specimens. Careful ion polishing will normally remove these surface films.

1.8.2 *Beam damage*

Organic materials such as polymers, proteins etc., together with inorganic salts, such as chlorides and bromides, and some silicate minerals, such as quartz, feldspar and clays, degrade during observation in the electron microscope because of damage due to ionization and/or heating from the electron beam. Often the material becomes amorphous (because the electron beam causes atomic bonds to be broken) and the diffraction pattern is lost (*Figure 1.13*); other materials transform to different compounds that can be identified by electron diffraction and X-ray analysis. Often the damage is so rapid that it is difficult to record a diffraction pattern under normal operating conditions.

There are several methods that you can use to minimize this type of beam damage.

- Operate at the highest possible kV. (The so-called **capture cross-section** for damage is *lower* the *higher* the kV.)
- Cool the specimen to the temperature of liquid nitrogen in a specially designed cold-stage. (The use of liquid helium appears only to be a marginal improvement over liquid nitrogen and is a lot more expensive!)
- Make preliminary adjustments such as setting the illumination conditions and focusing the pattern in a different area of the specimen from that you intend to use for recording. This can be done automatically on some instruments.
- Use a low beam-current and increase the exposure time accordingly. Or use an image intensifier.
- Use a short camera length, thus decreasing the exposure required. The pattern can be enlarged later.
- Encapsulate the specimen in a thin film of, for instance, carbon or calcium stearate. The protective effect is thought to be the result of the reduction in secondary damage. The diffusion of atomic and molecular species through the specimen is reduced and their recombination with damaged species is increased (see Misell and Brown, 1987).

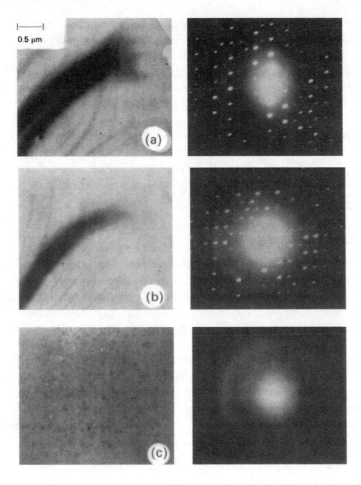

Figure 1.13. Effect of radiation damage on the alkali feldspar albite, $NaAlSi_3O_8$, at 200 kV on the image (left) and diffraction pattern (right). (a) After 1 min, (b) 5 min, (c) 12 min. The nominal beam flux was 3.2×10^{-2} cm^2. Note that the extinction contours visible in the image in (a) and (b) are no longer visible in the image in (c). The diffraction pattern present initially has disappeared after 12 min and is replaced by a faint, diffuse ring. Reproduced from Lorimer GW and Champness PE, *High Voltage Electron Microscopy*, 1974, with the permission of Academic Press.

The damage suffered by metals under the impact of the electron beam is of a different nature to that described above. This is called **knock-on** or **displacement damage** and occurs when electrons knock atoms out of their sites, so is directly related to the beam energy. Light metals such as Al and Mg suffer knock-on damage at 200 kV and, at 400 kV, metals with an atomic weight below that of Ti are damaged (see Williams and Carter, 1996). However, this damage is very slow and does not normally cause a problem for recording electron diffraction patterns.

1.9 Obtaining and recording a SAD pattern in the TEM

The complete procedure for obtaining a pattern is as follows.

- Select the required field of view; this must be within the SA range of magnification for FEI/Philips microscopes. Centre the beam and spread it to fill the field of view.
- Focus the image by using the **Objective focus** knob. Note that the image has minimum contrast when the objective aperture is removed and does not wobble when the **Wobbler** is switched on.
- Adjust the specimen height so that the specimen is in the eucentric plane (to ensure that the specimen is always at the same height for diffraction; i.e. the height for which the camera constant has been calibrated and the objective lens is working at its optimum setting).

  a. Use the beam stop to mark a prominent feature of the specimen and tilt the specimen by about 20° about the major tilt axis; the image will move across the screen.

  b. Adjust the specimen height using the *z* control (the thumbwheel on older Philips models) until the image returns to the centre of the screen.

  c. Re-zero the tilt, re-centre the image with the specimen translators and re-focus the image.

  d. Repeat steps a–c until there is minimum movement of the image.

  Note that on the CM and later series of FEI/Philips TEMs there is an **Autofocus** switch which sets the eucentric height; you then only have to refocus the image. Alternatively there is an '*α*-**Wobbler**' control that, when depressed, rocks the major tilt by ±15° about zero; the *z* control is then used to adjust the height until the image no longer appears to move about the centre of the screen.

- Remove the objective aperture.
- Focus the second condenser lens (the **Intensity** control on FEI/Philips machines and the **Condenser** control on JEOL machines) and centre the beam using the **Deflection** (FEI/Philips) or **Trans** (JEOL) control.
- Defocus the second condenser lens and refocus the image.
- Insert a diffraction aperture (you will have a choice of three or four sizes) and centre it over the region whose diffraction pattern you wish to see. If the aperture does not appear in the image, you will need to remove the aperture, decrease the magnification, spread the illumination and re-insert the largest aperture, centre it, and gradually increase the magnification again while keeping the aperture in view with the alignment controls.
- On JEOL microscopes depress the **Sam/Rock** switch and focus the aperture with the **Diffraction focus** knob.
- Select the diffraction mode.

- Choose the required camera length using the selector switch. (The camera length is changed by variation of the excitation of the intermediate and projector lenses.) *Figure 1.6* shows an example of a pattern recorded at two different camera lengths.
- Underfocus the second condenser lens (usually an anti-clockwise rotation of the knob) by several coarse steps to obtain a more parallel electron beam. The more C2 is defocused, the sharper the diffraction spots will be. Aperture sizes of 200 µm in the first condenser lens, C1, and 100 µm in C2 are typical values for imaging purposes, but if high beam coherence is required in diffraction, for instance for diffraction from materials with large atomic repeats such as biological crystals or polymers, smaller C1 and C2 apertures should be used. Reduction of the **Spot size** using the C1 control (increasing the excitation of C1) also increases the **coherence** of the electron beam (coherence is a difficult concept in physics, but is a way of defining how well the electrons in the beam are 'in step' or 'in phase' with each other; a parallel beam is more coherent than a convergent beam). It should be noted that all these measures for increasing the coherence of the beam will also reduce the intensity of the diffraction pattern and result in the need for longer exposures. Cooling the specimen to the temperature of liquid nitrogen also improves the sharpness of the spots because it reduces thermal-diffuse scattering in the specimen, the scattering produced by the thermal motion of the atoms.
- Focus the diffraction pattern using the **Diffraction focus** (FEI/ Philips) or **SA/HD diffraction camera length** control (JEOL). This focuses the diffraction lens accurately on the back focal-plane of the objective lens. (If the objective aperture is replaced, you should also see a sharp image of it).
- If necessary, the pattern can be centred on the screen using the **Diffraction point alignment** control (FEI/Philips) or the shift knob of the **Deflector–Projector** control (JEOL). In either case this procedure is achieved by deflection coils. It may be necessary to centre the pattern for each camera length.
- Correct any astigmatism of the diffraction lens (Section 1.11).
- If you want to take a photograph of the pattern, the exposure time needed will depend on the amount of defocus of C2, the type of specimen, etc. The exposure meter *cannot* be used for this purpose. The only way of finding the correct exposure is by experience or making a series of photographs at different exposure times. If the central beam is very intense compared with the other spots, the beam stop can be inserted and a double exposure taken. The first exposure includes the beam stop and the second is recorded without it for a length of time of about one 10th of the first exposure. An example of the use of the beam stop is shown in *Figure 1.6*. If you have a CCD (charge-coupled detector) imaging system or an image-plate system you will need to refer to the manufacturer's manual. Note that the CCD camera could be damaged when an intense diffraction pattern is focused on it.

1.10 Obtaining a CBED pattern

1.10.1 TEM mode

To form a microdiffraction pattern in the TEM mode the procedure is as follows.

- Make sure the specimen is in the eucentric plane and that the image is focused.
- Select a small C2 aperture (10–50 μm in diameter) and centre it.
- Decrease the spot size two or three clicks with the **Spot size** control (the excitation of C1 determines the spot size, which, in turn, determines the size of the area from which the diffraction pattern will be obtained). Remember that decreasing the spot size also decreases the electron-beam current at the specimen (it is proportional to α^2). You will probably have to correct the condenser astigmatism as you change the spot size.
- Focus the beam down onto the area of interest using the C2 lens control.
- Switch to the diffraction mode and remove the objective aperture. The pattern should now be visible. If the discs overlap, you will have to repeat the procedure from the second point using a smaller C2 aperture. Note: on the FX series and later JEOL machines the α-**Control** provides a continuous control of the convergence angle by changing the excitation of the auxiliary or mini-lens. A similar device will soon be available on the FEI/Philips Technai range.
- Use **Diffraction focus** to focus the image of the C2 aperture at the edge of one of the discs.
- You should check that the beam is accurately focused on the specimen. The procedure is to view the central beam through the binoculars and to overfocus (strengthen) either the C2 or the objective (**Focus**) lens so that an (out-of-focus) image is visible in the disc (*Figure 1.14a*). Then weaken the lens and the image will expand to higher magnifications until an inversion point is reached at exact focus (*Figure 1.14b*). At underfocus, an image will again be seen that is inverted with respect to the image at overfocus (*Figure 1.14c*).

1.10.2 STEM mode

The procedure is as follows:

- Obtain a focused image in STEM on the cathode-ray tube (see Keyse *et al.*, 1998 or Williams and Carter, 1996 for a description of the operation in STEM).
- Stop the beam scanning by selecting **Spot** on the STEM console.
- You may have to remove the STEM detector if it is above the TEM screen, or lower the TEM screen if the detector is below. Position the

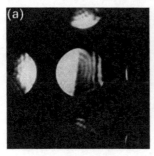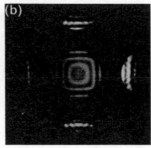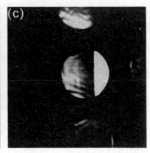

Figure 1.14. Procedure for focusing a CBED pattern by adjusting the strength of the objective or C2 lens. Under overfocus (a) or underfocus (c) conditions you see an image of the specimen in the central disc, but at exact focus (b) the disc contains non-spatial contrast. Reproduced from Williams DB and Carter ECB, *Transmission Electron Microscopy*, 1996, with the permission of Plenum Press, New York.

spot on the region of interest. A CBED pattern should be visible on the fluorescent screen of the TEM because the microscope is operated in the diffraction mode during STEM.

• You can change the camera length and focus the image of the C2 aperture as in the TEM modes. However, in TEMs in which the C2 lens is switched off in STEM operation, the diffraction pattern can only be focused using the objective lens.

1.11 Alignments that are important for diffraction

The routine alignments of the illumination and imaging systems that need to be are carried out before operation of the TEM are described by Chescoe and Goodhew (1990). One of these alignments that directly affects the diffraction pattern is correction of the **astigmatism** in the condenser lens. Astigmatism arises when the lens field is not perfectly symmetrical about the optic axis and results in a 'smearing out' of the image when it is not precisely in focus, and an unsharp image when apparently at focus. In the diffraction pattern, condenser astigmatism results in the diffraction spots being elliptical. Correction of astigmatism in the condenser lens is achieved by adjustment of the condenser stigmators, as described by Chescoe and Goodhew (1990).

Astigmatism in the diffraction lens also affects the diffraction pattern; in effect, it makes the magnification of the pattern on the screen or photograph unequal in different directions. 'Ring' patterns (see Section 2.5) show a series of concentric ellipses rather than circles. Any astigmatism is corrected as follows.

- Focus the image of a specimen within the SA range of magnifications and at the eucentric height.
- Remove the specimen.
- Select a small spot size with the C1 control and centre on the optic axis.
- Select the diffraction mode.
- Defocus C2 by several coarse clicks to obtain defocused illumination.
- Select the required camera length and centre the beam on the screen if necessary.
- Adjust the diffraction focus control until the image crossover is visible on the screen (*Figure 1.15a*)
- Adjust the diffraction stigmator controls until the crossover is a symmetrical three-pointed pattern as shown in *Figure 1.15b*.

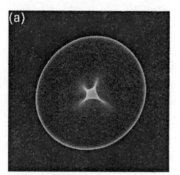 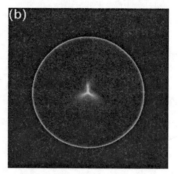

Figure 1.15. Correction of astigmatism in the diffraction lens. (a) Uncorrected; (b) corrected.

Note that not all microscopes have the requisite stigmators to allow this correction.

2 The reciprocal lattice and Bragg's Law

In Chapter 1 we saw how a diffraction pattern is formed by a simple lens in its back focal-plane. We also noted that the pattern is related in a reciprocal fashion to the object from which it comes: a grating, or other periodic object, of spacing d produces a row of spots perpendicular to the grating with a spacing proportional to $1/d$. This reciprocal relationship may, as we shall see, be extended to three-dimensional gratings, i.e. crystals, to give what is known as the reciprocal lattice.

If you are not familiar with the crystal systems, Miller indices and zone directions, it is advisable to read Appendix A before tackling this chapter.

2.1 Electrons as waves

It is well known that light and electrons have both wave-like and particle-like characters. The relationship between their two characters is shown by Planck's equation:

$$E = hc / \lambda$$

where E is the energy of the photon or 'particle', c is the velocity of light ($\sim 3 \times 10^8$ ms^{-1}), λ is the wavelength (see below) and h is Planck's constant (6.6256×10^{-34} Js). However, it is the wave-like nature of electrons that we need to consider if we want to understand how they interact with crystals to produce diffraction patterns. Electrons and light waves, like water waves, are transverse waves. That means that the disturbance or **displacement** they produce (in the electromagnetic field in the case of light) is perpendicular to their direction of travel. *Figure 2.1* shows the sinusoidal profile of a light or electron wave in which the distance x travelled by the wave from its source is plotted against the displacement y. The maximum value of the displacement is called the **amplitude**, A, of the wave (the intensity, I, of the wave is simply proportional to A^2). The distance between successive maxima (or any two other equivalent points) is the wavelength, λ.

The wavelength of visible light waves varies between 390 nm (violet light) and 707 nm (red light) (1 nm = 10^{-9} m = 10 Å). The wavelength of

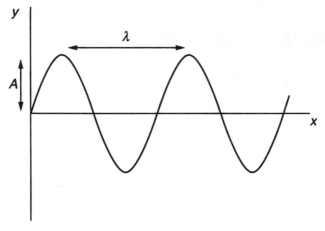

Figure 2.1. The displacement *y* of a wave with distance *x*. *A* is its amplitude and λ is its wavelength.

electrons depends on their velocity, v, and consequently varies with the accelerating voltage, V, used. The relationship between the velocity and wavelength is given by the de Broglie equation:

$$\lambda = h/mv$$

where m is the relativistic mass of the electron. The energy given to the electron is:

$$eV = \frac{1}{2}mv^2$$

where e is the charge on an electron. Hence:

$$\lambda = \frac{h}{\sqrt{2meV}}.$$

The wavelengths of electrons calculated from this equation for common values of V are shown in Appendix D. Note that the wavelength for electrons accelerated at 200 kV (2.508 pm) is five orders of magnitude smaller than the wavelength of visible light and is about two orders of magnitude smaller than the size of a typical atom. This is the reason why electrons are diffracted by the atoms in a crystal, but light waves are not; light cannot 'see' the atoms because it is a fundamental law of physics that radiation cannot detect detail in an object that is less than about half its wavelength.

2.2 Interference of waves

We must now consider how two or more diffracted waves interact with each other. For waves of the same wavelength vibrating in the same plane, which is what we are mostly concerned with in electron diffraction, the rules that govern their interaction or **interference** are very simple; the **principle of superposition** states that the resultant wave can be obtained by the summation of the component waves. *Figure 2.2a* shows two waves with the same wavelength and amplitude, A, that are **in phase**, i.e. their maxima and minima coincide. The resultant wave, shown on the right, is a wave of the same wavelength, but with twice the amplitude, $2A$. The waves are said to show **constructive interference** and the resultant amplitude is the maximum that could be achieved by the interference of the two waves. As the wave profiles are identical, if they are shifted by λ, constructive interference will also be obtained if the path difference, Δ, between the two waves is a whole number of wavelengths. On the other hand, if the maxima of one wave coincide with the minima of the other (*Figure 2.2b*), **destructive interference** occurs and the resultant displacement and the intensity are zero. In this case, the **path difference**, Δ, is $\lambda/2$ or any odd multiple of $\lambda/2$. Waves

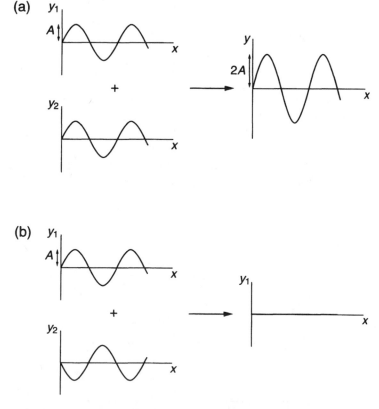

Figure 2.2. The superposition of two waves of the same wavelength and amplitude, A. (a) The waves are **in phase**; constructive interference; path difference $n\lambda$, where n is an integer. The resultant amplitude is $2A$. (b) The waves are **out of phase**; destructive interference; path difference $(n + 1/2)\lambda$.

with a path difference intermediate between those shown in *Figure 2.2a* and *b* will obviously produce a resultant amplitude between zero and 2A.

2.3 Bragg's Law

Now that we understand how waves interfere we are in a position to look into how crystals diffract electrons. The situation is more complex than with the diffraction from the one- or two-dimensional gratings and grids that we considered in Chapter 1 because crystals are three dimensional and can contain many hundreds or thousands of atoms in their unit cells or repeat units. However, the equation that is most useful for understanding the geometry of the diffraction process in crystals is remarkably simple.

W.L. Bragg envisaged diffraction in crystals in terms of reflections from planes. For this reason, we refer to the **diffraction** pattern being formed by **reflections** from sets of planes in the crystal. *Figure 2.3* shows three successive crystal lattice planes with Miller indices (hkl) and spacing d_{hkl}, in a projection in which we are looking along the planes, so that they project as lines. A beam of electrons of wavelength λ is impinging on the planes at a glancing angle θ and being reflected by the planes, also at angle θ. For the incident beam on the left, X and G are equivalent points for the rays impinging on successive planes P and Q; likewise, X and H are equivalent points on the rays that are reflected. The path difference between rays reflected by P and Q is therefore GY + YH. For there to be constructive interference and a maximum in the intensity of the reflected beam, this distance must be equal to a whole number of wavelengths, λ (in the figure it is actually shown to be equal to λ). Looking at the geometry of *Figure 2.3* we can see that, since angle GXA equals 90° and

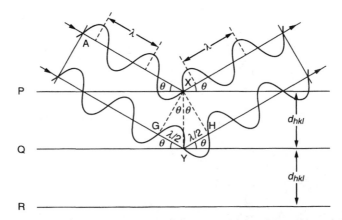

Figure 2.3. The Bragg equation. P, Q and R are three successive lattice planes, (hkl) in a crystal. The incoming radiation, wavelength λ, is incident on the planes, spacing d_{hkl} at a glancing angle θ and is reflected by the planes at angle θ. The path difference between the rays reflected by planes P and Q is λ, which, by geometry, is equal to $2d\sin\theta$.

angle AXP is θ, angle PXG equals $90-\theta$. As PXY is 90°, the angle GXY is equal to θ. In the right-angled triangle XGY:

$$\sin \theta = GY/d, \therefore GY = d \sin \theta$$

As GY = YH, the total path difference is $2d \sin \theta$, and for constructive interference:

$$2d_{hkl} \sin \theta = n\lambda$$

where n is an integer representing one, two, three, etc. wavelength path differences. This is **Bragg's Law**. When this equation is obeyed, θ is known as the **Bragg angle**.

Although Bragg's approach considers the radiation to be reflected from crystal planes, there is a very important difference between these 'reflections' and the reflections of light from a mirror: light is reflected whatever the angle of incidence, whereas radiation incident on a set of crystal planes is only reflected or diffracted at the specific angles given by Bragg's Law. This difference is a consequence of the fact that the crystal is three dimensional and specifically that, for rays reflected from the top plane (as with light), the path difference is always zero and the waves always interfere constructively.

In light optics, when $n = 1$ in Bragg's Law we refer to the first-order diffracted beam, when $n = 2$ to the second-order diffracted beam, etc. For diffraction from crystal planes, however, we conventionally write:

$$2d_{hkl} \sin \theta = \lambda, \qquad \text{i.e. } n = 1.$$

For example, rather than referring to the second-order reflection from the (100) planes, we consider that the radiation is diffracted by planes of half the spacing, i.e. the (200) planes (*Figure 2.4*). (The Miller indices are reciprocals of the fractional intercepts of the planes on the unit-cell axes, Appendix A.) In terms of Bragg's equation we can write:

$$2\left(\frac{d_{hkl}}{n}\right)\sin \theta = \lambda = 2d_{nh\ nk\ nl} \sin \theta.$$

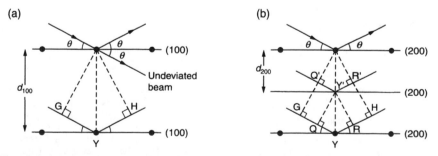

Figure 2.4. (a) The Bragg condition for the second-order reflection from the (100) planes. If the path difference, GY + YH, between rays reflected by successive planes is 2λ, the Bragg equation can be written: $2d_{100} \sin \theta = 2\lambda$; (b) The same condition, but considered as the *first-order* reflection from the (200) planes. If GY + YH = 2λ, then QY + YR = λ. This distance is clearly equal to Q'R' + Y'R', the path difference for the (200) set of planes.

It is clearly more convenient to refer to the '200 reflection' than to 'the second-order reflection from the (100) planes'. Notice that, when we refer to reflections from planes, we omit the brackets that we use when we refer to the planes themselves; i.e. the (*hkl*) planes give rise to the *hkl* reflection. (Note that multiple indices such as 200, 300, etc. are sometimes called **Laue indices** to distiguish them from Miller indices, which have no common factors.)

The angle that the diffracted beam makes with the incident (undeviated) beam is 2θ (*Figure 2.4a*). It follows that the distance from the diffraction spot to the centre of the diffraction pattern is proportional to 2θ. But from Bragg's Law:

$$\sin\,\theta \sim \theta = \lambda/2d$$

(because θ is small for electrons: 0.14° or 2.5×10^{-3} radians for a crystal spacing of 0.5 nm and 200 kV electrons). In this equation, we can see once again the reciprocal relationship between the spacing of the diffraction spots and the spacing of the crystal planes from which they come.

Each set of diffracting planes (*hkl*), spacing d_{hkl}, produces a spot in the diffraction pattern at a distance R from the centre and in a direction perpendicular to the planes. From *Figure 2.5* we can see that:

$$R/L = \tan\,2\theta \sim 2\theta\ \text{(because θ is small)}$$

where L (called the camera length) is the distance of the specimen from the screen or film on which the diffraction pattern is displayed. In an electron microscope, this is *not* a physical distance, but a 'projected' distance which varies with the setting of the imaging lenses. We can combine the above equation with Bragg's Law:

$$\lambda = 2d\sin\,\theta \sim 2d\theta,\ \therefore\ dR = \lambda L.$$

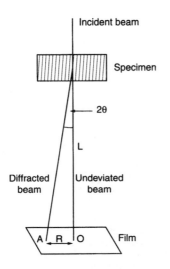

Incident beam

Specimen

2θ

L

Diffracted beam

Undeviated beam

A R O   Film

Figure 2.5. Schematic diagram showing the geometry of the formation of an electron diffraction pattern. On the film, the spot A occurs at a distance *R* from the central spot, O (that formed by the direct beam) in a direction perpendicular to the planes. *L* is known as the camera length.

λL is known as the **camera constant** and relates the distance of a spot from the origin of the diffraction pattern to the d-spacing of the set of planes from which it comes. The camera constant can be determined using a standard (see Section 2.10), and hence the d-spacing corresponding to a reflection can be found once R has been measured.

2.4 The reciprocal lattice

The angles in *Figures 2.3–2.5* have been greatly exaggerated and, in practice, the electron beam is diffracted from crystal planes that are almost parallel to the electron beam. (For a crystal spacing of 0.5 nm the Bragg angle for 200 kV electrons is 0.0025 radians $= 0.14°$.) If there are several planes parallel to the electron beam, they will each produce a diffraction spot in the electron microscope at a distance $\lambda L/d$ from the centre of the diffraction pattern and along the direction of the normal to the plane (*Figure 2.6*). The angle ϕ that any two spots subtend at the spot from the undeviated beam is equal to the angle between the normals to the planes.

When a number of planes are parallel to a single direction, they are said to constitute a **zone** and the common direction is the **zone axis** or **zone direction** (Appendix A). Thus, when the incident electron beam is parallel to a zone axis in the crystal, the diffraction pattern consists of a lattice of spots that bears a reciprocal relationship to the real (direct) lattice of the crystal – the **reciprocal lattice**. Such diffraction patterns are normally identified by the zone axis to which the electron beam is parallel. It follows that the indices hkl of all the spots in the pattern are

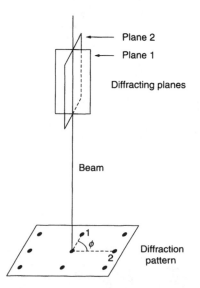

Figure 2.6. Schematic diagram of the formation of a diffraction pattern from a single crystal in the electron microscope. The beam is parallel to a zone axis and two of the planes in that zone are shown. They produce spots 1 and 2 in the diffraction pattern, and the angle ϕ that the spots subtend at the spot from direct beam is equal to the angle between the normals to the planes themselves.

related to the indices of the zone axis [*UVW*] to which the beam is parallel by the Weiss zone law (or zone equation): $hU + kV + lW = 0$ (see Appendix A).

We can define the reciprocal lattice as a lattice of spots, each of which represents a set of planes (*hkl*) of spacing *d* in the real lattice. Each spot lies at a distance $g = 1/d$, the reciprocal lattice vector, from the origin and is aligned in the direction perpendicular to the planes. (Strictly the vector is **g** and $g = |\mathbf{g}|$ is its modulus). This definition allows us to construct the reciprocal lattice for any crystal whose unit cell parameters we know (*d*-spacings of planes can be calculated using the equations in Appendix C).

Let us look at the reciprocal lattice for the [001] zone of an orthorhombic crystal. In this system, the axes are at right angles, but *a* $\neq$ *b* $\neq$ *c* (see *Table A.1*). In *Figure 2.7a* the real lattice is projected onto the (001) plane. It contains the x and y axes, and some of the members of the set of planes (100), (010) and (110) have been drawn in projection as lines. The reciprocal lattice spot for the (100) set of planes will lie along the normal to the (100) planes at a distance:

$$1/d_{100} = 1/a$$

from the central spot, 000, which comes from the undeviated beam, as shown in *Figure 2.7b*. The 200 spot will lie in the same direction as 100, but at a distance of:

$$1/d_{200} = 2/a.$$

In fact, there will be a row of spots along this direction, consisting of the

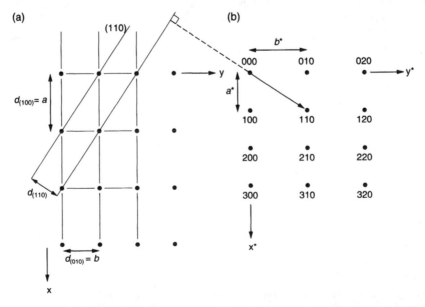

Figure 2.7. Sections of the direct and reciprocal lattices for an orthorhombic crystal. (a) Projection of the direct lattice onto the (001) plane; (b) the corresponding [001] reciprocal lattice.

100, 200, 300, 400.... reflections. This direction is called the x* axis (a star signifying a parameter in the reciprocal lattice) and the repeat along it is called $a*$ (the vector is **a***).

In the same way the (0k0) series of planes will produce a row of spots perpendicular to the (010) planes with a spacing:

$$b* = 1/d_{010} = 1/b.$$

This is the y* axis of the reciprocal lattice. The (110) set of planes will produce a spot along their normal (*Figure 2.7b*) at a distance of:

$$g_{110} = \frac{1}{d_{110}} = \sqrt{\frac{1}{a^2} + \frac{1}{b^2}}$$

(see Appendix C). In fact, you will find that the reciprocal lattice spot for any plane that you draw in *Figure 2.7a* is along the normal to that plane in *Figure 2.7b*.

The three-dimensional reciprocal lattice for an orthorhombic crystal is shown in *Figure 2.8*. The third dimension z* is perpendicular to the (001) planes of the real lattice and the repeat along the axis is:

$$c* = 1/d_{001} = 1/c.$$

The following relationship holds for *any* crystal system, simply by vector addition:

$$\mathbf{g}_{hkl} = h\mathbf{a}* + k\mathbf{b}* + l\mathbf{c}* = \mathbf{g}_{h00} + \mathbf{g}_{0k0} + \mathbf{g}_{00l}$$

In other words, a reciprocal lattice point hkl is found by going h units along **a***, k units along **b*** and l units along **c***. For example, in *Figure 2.8* the $\overline{11}2$ reflection lies one unit along the negative x* direction (-**a***), one unit along the negative y* direction (-**b***) and two units along the positive z* direction (2**c***).

Also, for any crystal system:

$$a* = 1/d_{100}; \quad b* = 1/d_{010}; \quad c* = 1/d_{001}.$$

Thus, **a*** is always perpendicular to the plane (100) even when **a** is not. Also:

- x* is always perpendicular to y and z;
- y* is always perpendicular to z and x;
- z* is always perpendicular to x and y;
- $\alpha*$ is the angle between the x* and y* axes and is equal to the angle between the normals to the (010) and (001) planes;
- $\beta*$ is the angle between the x* and z* axes and is equal to the angle between the normals to the (100) and (001) planes;
- $\gamma*$ is the angle between the x* and y* axes and is equal to the angle between the normals to the (100) and (010) planes.

The mathematical definitions of the reciprocal lattice in vector notation are given in Appendix E.

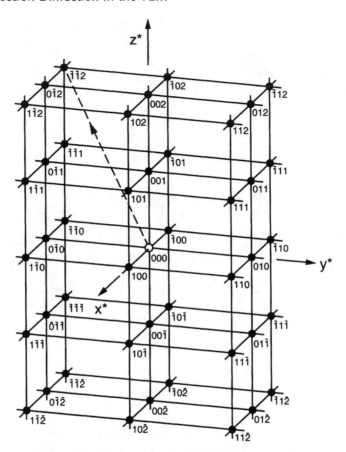

Figure 2.8. Part of the three-dimensional reciprocal lattice of an orthorhombic crystal. The x*, y* and z* axes are at right angles. The **g**-vector for reflection $\overline{1}12$ is shown.

Although the 'real' or direct lattice and the reciprocal lattice look geometrically similar, there is one vital difference between them. Each node of the real lattice is identical and any one can be taken as the origin; in the reciprocal lattice, however, each node is distinct as it represents a different set of planes. The origin of the reciprocal lattice is 000 and represents the undeviated beam.

The [010] reciprocal lattice for a monoclinic crystal is shown in *Figure 2.9*, together with the (010) projection of the real lattice. For this system: $a \neq b \neq c$, $\alpha = \gamma = 90°$, $\beta > 90°$. Thus, the x- and z-axes are shown with an obtuse angle β between them. In the reciprocal lattice, the spot 100 will lie along the normal to the (100) planes at a distance from the origin of:

$$1/d_{100} = 1/(a \sin \beta) = a^* \text{ or } |\mathbf{a}^*|$$

The sin β term arises because the x and y axes are not at right angles, i.e. $d_{100} = a \sin \beta$.

Similarly, the 001 spot will lie along the normal to the (001) planes at a distance:

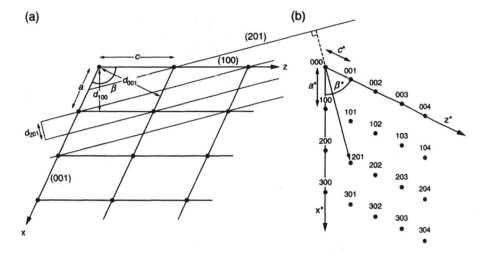

Figure 2.9. Sections of the direct and reciprocal lattice for a monoclinic crystal. (a) Projection of the direct lattice onto the (010) plane; (b) the corresponding [010] reciprocal lattice. Some of the (201) planes are draw in (a). In (b) the corresponding reciprocal lattice point is found in a direction normal to the planes.

$$1/d_{001} = 1/c \sin \beta = c^*.$$

The resultant reciprocal lattice is a parallelogram in which the angle between x* and z* is β* which is equal to $180 - \beta$.

As a further example, we will look at the [001] reciprocal lattice of a hexagonal crystal. This crystal system also has axes that are not orthogonal: the x- and y-axes are at 120° (the γ angle) and there is a third axis, u, in the (001) plane that is symmetrically equivalent to x and y and at 120° to them (*Figure 2.10a*). For this reason, as explained in Appendix A, we normally use *four* indices (the **Miller–Bravais indices**, order x, y, u, z) to describe hexagonal (and trigonal) crystal planes and the diffraction spots from them.

Figure 2.10b shows the [001] reciprocal lattice section through the origin with a number of the reflections indexed. Here γ* is 60° (180–γ) and:

$$a^* = b^* = \frac{1}{d_{10\bar{1}0}} = \frac{1}{a\sin 60°} .$$

2.4.1 The 180° ambiguity in indexing the diffraction pattern

Every diffraction pattern from a zone axis can be indexed using a positive or negative set of indices, i.e. a particular spot may be indexed as *hkl* or as $\overline{hkl}$ (*Figure 2.11a, b*). If you need to find the orientation of the crystal producing the diffraction pattern with respect to another crystal, two

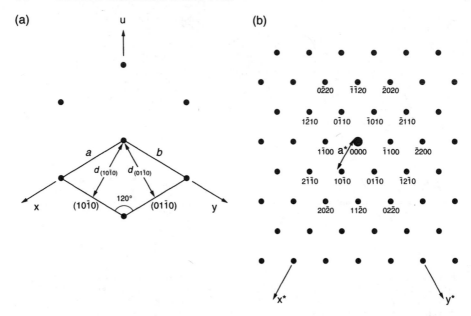

Figure 2.10. Sections of the direct and reciprocal lattice for a hexagonal crystal. (a) Projection of the direct lattice onto the (0001) plane; (b) the corresponding [001] = [0001] reciprocal lattice.

different answers will be obtained for the two different methods of indexing the pattern because they are equivalent to a rotation of the crystal by 180° about the zone axis (*Figure 2.11c, d*). This ambiguity is irrelevant to most applications of electron diffraction, but needs to be resolved for certain analyses; for example, the determination of the burgers vectors of dislocations. Edington (1975) explains in detail how the ambiguity may be removed.

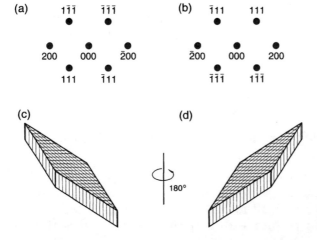

Figure 2.11. The influence of reversing the indices on the diffraction pattern in (a) and (b) on the effective position of the specimen in (c) and (d). Adapted from Edington JW, *Electron Diffraction in the Electron Microscope*, 1975, with permission of Macmillan Press.

2.5 Diffraction patterns from polycrystalline materials

Up to now we have dealt with diffraction from single crystals. If the area of the specimen selected by the diffraction aperture contains crystals in several orientations, the diffraction pattern will consist of the sum of the individual patterns. In the case where the specimen consists of very many crystals of random orientation, the spots are so close together that they fall on a series of continuous rings (*Figure 2.12*) and form a so-called **powder diffraction pattern**. Each ring comes from one set of planes of spacing d and the radius R of the ring is equal to $\lambda L/d$. There will be one ring from *each* set of planes in the structure that satisfies Bragg's Law. If λL is known, the d-spacings corresponding to the rings can be calculated.

2.6 Diffraction patterns from amorphous materials

A diffuse ring pattern indicates that the specimen is amorphous. Such patterns arise from the support film, e.g. carbon, or areas of the specimen that have no crystallinity. The d-spacings of the diffuse rings roughly correspond to the average nearest and next-nearest atomic spacings in the material. An example of a diffraction pattern from an amorphous phase is shown in *Figure 2.13* (bottom right). In this case, the lack of crystallinity is the result of radiation damage by the electron beam.

Figure 2.12. Electron diffraction pattern from polycrystalline gold: the Miller indices of the planes contributing to the diffraction rings are shown.

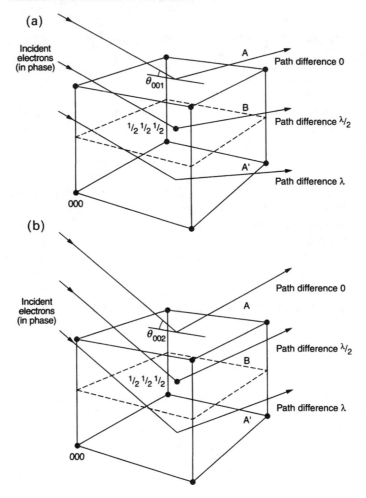

Figure 2.13. The phase relationships for reflections by lattice planes for (a) (001) and (b) (002) planes in a crystal with a body-centred lattice. The dots represent lattice points. The path differences given are relative to the reflected ray from plane A.

2.7 Systematic absences

If the lattice of the crystal is not primitive, certain reflections that would otherwise be present in the diffraction pattern have zero intensity; in other words, they are absent. You must be aware of the rules that govern when such reflections are **systematically absent** or **forbidden**. Otherwise you may not be able to index experimental diffraction patterns, i.e. to identify the zone axis whose planes are contributing to a pattern and to identify the Miller indices of the individual spots and of the planes from which they come.

The origin of systematically absent reflections will be covered quantitatively in Chapter 5, but here it will be shown how the absences for a body-centred lattice can be derived by taking a more qualitative approach. *Figure 2.13* shows the lattice points at the corners and body centre of the unit cell (remember that the lattice points may represent a large number of atoms that form the repeat unit or motif of the structure). In *Figure 2.13(a)*, a beam of electrons is incident on the (001) set of planes at the Bragg angle θ_{001}. That means that the reflected electrons from successive (001) planes A and A' are in phase and have a path difference of λ. However, there is an identical plane, B, half way between the A and A' planes, and electrons reflected from it will be out-of-phase with a path difference of $\lambda/2$ compared with the reflection from plane A. Therefore, destructive interference will occur and the resultant intensity will be zero.

The planes A, A' and B make up part of the set (002) (half the spacing of the (001) planes); in *Figure 2.13b*, the electrons are incident on these planes at the Bragg angle θ_{002}; reflections from successive planes of the set will have a path difference of λ and be in phase. The 002 reflection will therefore be present. Similar arguments show that reflections 400, 600, etc. will be present, but 300, 500, etc. will be absent; 110, 220, 330.... reflections will all be present, but 111, 333, 555.... will be absent. Consideration of these absences shows that, for this lattice type, if $h + k + l$ is odd, then the reflection is absent. *Table 2.1* shows the conditions for reflections to occur for all the possible lattice types.

Figure 2.12 is a powder diffraction pattern from gold. Because we know that gold is a face-centred cubic we can index the powder rings. For cubic crystals:

$$d_{hkl} = \frac{a}{\sqrt{h^2 + k^2 + l^2}}$$

Table 2.1. Conditions for reflection for different lattice types

Lattice type	Coordinates of lattice points	Conditions for reflection
Primitive, P	0,0,0	None, all present
Body centred, I	0,0,0; 1/2, 1/2, 1/2	$h + k + l = 2n$ (even)
All face centred, F	0,0,0; 0, 1/2, 1/2; 1/2, 0, 1/2 ; 1/2, 1/2, 0	h, k, l all even or all odd
C-face centred, C	0,0,0; 1/2, 1/2, 0	$h + k = 2n$
B-face centred, B	0,0,0; 1/2, 0, 1/2	$h + l = 2n$
A-face centred, A	0,0,0; 0, 1/2, 1/2	$k + l = 2n$
*Rhombohedral, R (obverse)	0,0,0; 2/3, 1/3, 1/3; 1/3, 2/3, 2/3	$h - k - l = 3n$
*Rhombohedral, R (reverse)	0,0,0; 2/3, 1/3, 2/3; 1/3, 2/3, 1/3	$h - k + l = 3n$

Conditions for non-standard lattice types A and B, (100) and (010) faces centred respectively, are given for completeness.
*When indexed using hexagonal indices. If rhombohedral axes are used (which is rare) the lattice is primitive. The inverse and reverse rhombohedral cells differ only by a rotation of 180° about the z-axis.

(Appendix C) and we also know from Bragg's Law that:

$$d_{hkl} = \frac{\lambda}{2 \sin \theta} \sim \frac{\lambda}{2\theta}.$$

Therefore:

$$2\theta \propto \sqrt{h^2 + k^2 + l^2}$$

Thus, for cubic crystals the indices of the rings in a powder diffraction pattern have progressively increasing values of $N = h^2 + k^2 + l^2$ from the centre of the pattern outwards. So the first ring will have $N = 1$, 2 or 3, depending on whether the lattice is P, I or F (*Table 2.2*). The first four rings in the Au pattern in *Figure 2.12* are therefore: 111, 200, 311 and 222.

2.8 Standard patterns

2.8.1 *Cubic patterns*

For cubic crystals:

$$d_{hkl} = \frac{a}{\sqrt{h^2 + k^2 + l^2}}$$

Table 2.2. Possible $h^2 + k^2 + l^2$ values for the cubic lattice types

$h^2 + k^2 + l^2$ = N	P hkl	I hkl	F hkl
1	100		
2	110	110	
3	111		111
4	200	200	200
5	210		
6	211	211	
8	220	220	220
9	300, 221		
10	310	310	
11	311		311
12	222	222	222
13	320		
14	321		
16	400	400	400
17	322, 410		
18	330, 411	330, 411	

Note that $h^2 + k^2 + l^2$ cannot equal 7 or 15 because no sums of three squared integers add up to these values. Remember that, in the cubic system, because $a = b = c$, you can permute the indices to find planes of the same form. For example, 200 is of the same form as $\overline{2}00$, 020, $0\overline{2}0$, 002 and $00\overline{2}$ and therefore the d-spacings of all these planes are the same. Also, although they are not planes of the same form, the (300) and (221) planes have the same d-spacing because $h^2 + k^2 + l^2$ is equal to 9 for both of them.

and therefore the ratio of any two d-spacings, d_1/d_2, and hence the ratio of any two g or R values, in a cubic diffraction pattern is independent of the lattice parameter, a. This means that the diffraction pattern for a particular zone for any cubic crystal with the same lattice type will always look the same, except for scale. *Figure 2.14* shows four high-symmetry (low-index) reciprocal lattice planes for each of the three different lattice types that occur for cubic crystals. Although the diffraction patterns of prominent zones can (with practice) be identified by comparing them by eye with the standard patterns, you may find it helpful to measure the patterns to be sure of an identification. For this

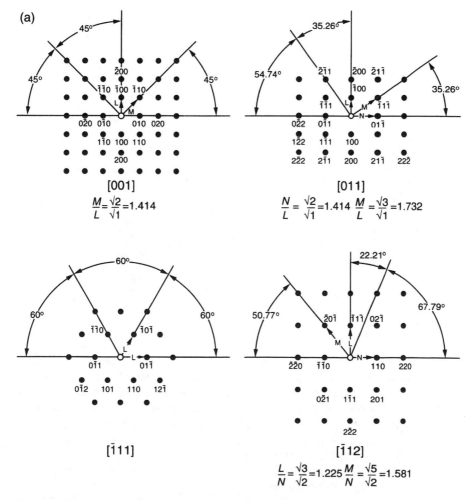

Figure 2.14. Four standard diffraction patterns for each of the three lattice types found in cubic crystals. They are drawn for the same lattice parameter and camera constant. Ratios of the principal spot spacings and the angles between some of the **g**-vectors are given. (a) The primitive cubic lattice. All possible reflections are present. (b) The body-centred cubic lattice. The reflections all have $h + k + l$ even. (c) The face-centred cubic lattice. The reflections have h, k and l all odd or all even.

(b)

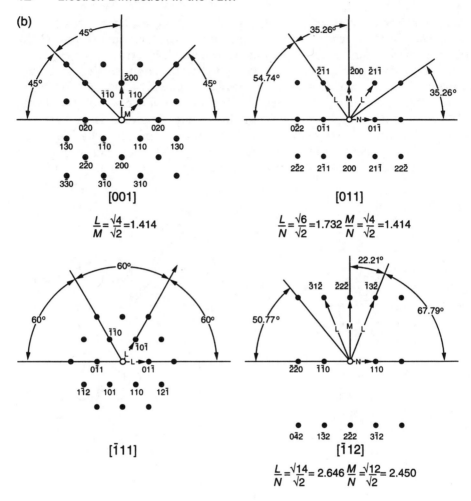

Figure 2.14. (b)

reason, the ratios of the principal spot spacings and some of the angles between them are given. Another useful thing to remember is that, if you know the camera constant (we will see in the Section 2.10 how to measure this), you can work out d for the spots, using the equation: $d = \lambda L/R$. The value obtained can then be compared with calculated or tabulated values of the material concerned. Alternatively, once you have identified the zone axis of a diffraction pattern from an unknown cubic material and indexed the spots, you can find the value of the a lattice parameter from the d-values using the equation above.

2.8.2 Hexagonal close-packed patterns

A number of pure metals, for instance Mg, Ti, Zr, have the hexagonal close-packed structure (see Hammond, 2001, for details of the structure).

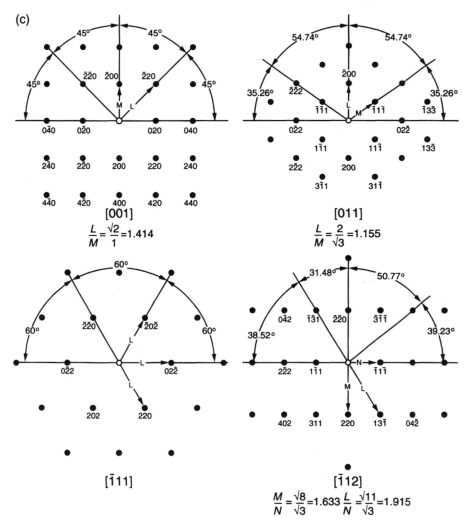

Figure 2.14. (c)

Because the ratio of the unit cell axes c/a is approximately a constant for this structure, the diffraction pattern from a particular zone will, like those from cubic crystals, look the same, apart from the scale. Standard patterns of hexagonal close-packed structures are therefore published (*Figure 2.15*). You should note however that, because the hexagonal system has a lower symmetry than the cubic system, the hexagonal close-packed structure gives rise to many more distinct reciprocal lattice planes. This means that the reciprocal lattice sections cannot always be uniquely identified as a result of a quick inspection, and measurement of the pattern may be essential. An example of this ambiguity is shown in *Figure 2.15b* and *d*: the $[120] = [01\bar{1}0]$ and $[122] = [01\bar{1}2]$ reciprocal lattices are very similar, both in shape and size.

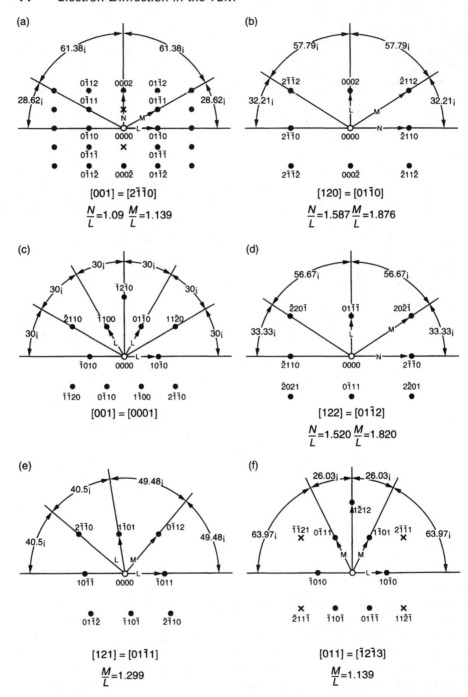

Figure 2.15. Six standard diffraction patterns for the hexagonal close-packed structure. Ratios of the principal spot spacings and the angles between the **g**-vectors are given. The patterns are drawn for a *c/a* ratio of 1.633. The positions of the spots will differ from those shown with a change of *c/a*. The positions marked x show reflections that are 'forbidden', i.e. they are systematically absent (see text for details).

The positions marked x in the schematic patterns of *Figure 2.15* show reflections that are 'forbidden', i.e. they are systematically absent, not because of the lattice type (which is primitive), but either because of the presence of a symmetry element known as a c-glide plane (see Appendix B) parallel to $\{1\bar{1}00\}$ in the structure or because all the atoms are in positions of high symmetry, known as **special positions,** in the unit cell. This kind of absence is covered in Chapter 5 and the reason need not bother us at this point. These 'forbidden' reflections reappear in the diffraction pattern of any sample of normal thickness by a process known as **double diffraction** (see Section 7.1), so you would be unaware that they were 'forbidden'! You may notice that, for the $[120] = [01\bar{1}0]$ pattern, half the reflections are missing and are not marked x, for instance 0001, 0003, $\bar{2}111$, etc. These reflections are also forbidden for the same reason as those marked x in the other patterns but, in this case, they do not reappear by double diffraction.

2.8.3 *Patterns from the diamond structure*

Diamond, silicon and germanium have face-centred cubic lattices and so show systematic absences in their diffraction patterns when h, k and l are not all odd or all even. However, like the hexagonal close-packed structure, there are two atoms per lattice point and both are in special positions in the unit cell. This leads to extra conditions for reflections to be present: $h + k + l$ is odd or $h + k + l = 4n$, where n is an integer (*Figure 2.16*).

2.9 Computer indexing of electron diffraction patterns

For any crystal that is not cubic or hexagonal close-packed, indexing of the diffraction patterns is best done with the aid of a computer. There are several programs that generate reciprocal lattice sections (e.g. *Figure 2.17*). Some will also index experimental patterns for you if you input the d-spacings for two spots and the angle between them. Some programs can be used 'on-line' at the microscope if the instrument has facilities for measuring d-spacings; you may even be able to use the program to tilt the stage of the TEM to the zone axis you want once two other zone axes have been identified! See Appendix F for a list of computer programs concerned with electron diffraction.

2.9.1 *Input data*

The input data for the computer program may simply be the cell parameters and the lattice type, as in the program used to produce

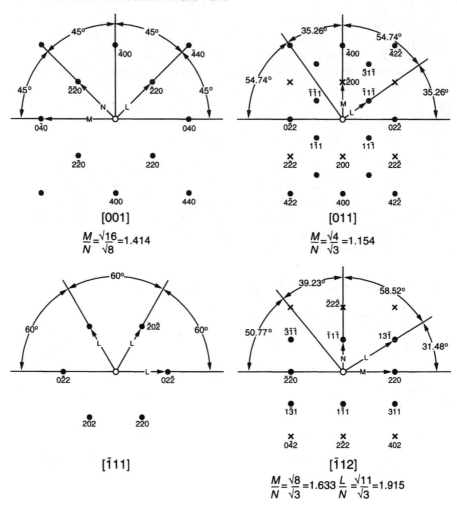

Figure 2.16. Four standard diffraction patterns for the diamond cubic structure. They are drawn for the same lattice parameter and camera constant as *Figure 2.14* . Ratios of the principal spot spacings and the angles between some of the **g**-vectors are given. In addition to the absences for the face-centred lattice, additional absences occur because all the atoms are in special positions in the unit cell. The extra conditions for reflections being present are: $h + k + l$ is odd or $h + k + l = 4n$, where n is an integer. Many of these reflections appear in the diffraction pattern by double diffraction, as shown in the figure by crosses.

Figure 2.17, but other programs require the space group and the positions of the atoms to be given. The most comprehensive source of data for the cell parameters, lattice type and space group is the Powder Diffraction File. An example of a card from the file is shown in *Figure 2.18*. The middle box on the left-hand side gives the crystallographic data. The first line gives the crystal system (Sys.), tetragonal in this case, followed by the space group (S.G.). You can learn more about space groups in Appendix B, but all you need to know at this stage is that the first letter of the symbol is the lattice type; P, or primitive, in this case.

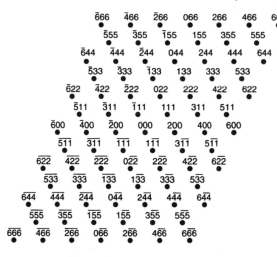

Figure 2.17. An adaptation of a diffraction pattern generated by computer. The [0$\bar{1}$1] zone-axis pattern from plagioclase feldspar which is C-face-centred triclinic with $a = 0.8169$, $b = 1.2851$, $c = 0.7124$ nm, $\alpha = 93.63$, $\beta = 116.4$, $\gamma = 89.46°$. Note that the reflections have $h + k$ even.

The second line in the middle box gives the unit-cell parameters, a, b, c, in Ångströms (1 Å = 0.1 nm). Where a value is left blank it is implied that it is equal to the one preceding it. In *Figure 2.18*, for example, $a = b$ as rutile is tetragonal. In the third line the inter-axial angles α, β and γ are shown. In the example these values are not given because, for all tetragonal crystals: $\alpha = \beta = \gamma = 90°$. As another example, for a hexagonal crystal the values of α, β and γ would also not be given because, for all hexagonal crystals, $\alpha = \beta = 90°$ and $\gamma = 120°$. The relationship between

21-1276							
d	3.25	1.69	2.49	3.25	TiO₂		★
I/I₁	100	60	50	100	Titanium Oxide	(Rutile)	

Rad. CuKα₁ λ 1.54056 Filter Mono. Dia.			
Cut off I/I₁ Diffractometer			
Ref. National Bureau of Standards, Mono. 25, Sec. 7 (1969)			
Sys. Tetragonal S.G. P4₂/mnm (136)			
a₀ 4.5933 b₀ c₀ 2.9592 A C 0.6442			
α β γ Z 2 Dx 4.250			
Ref. Ibid.			
εα nωβ εγ Sign			
2V D mp Color			
Ref.			
No impurity over 0.001%			

d A	I/I₁	hkl	d A	I/I₁	hkl
3.25	100	110	1.0425	6	411
2.487	50	101	1.0364	6	312
2.297	8	200	1.0271	4	420
2.188	25	111	0.9703	2	421
2.054	10	210	.9644	2	103
1.6874	60	211	.9438	2	113
1.6237	20	220	.9072	4	402
1.4797	10	002	.9009	4	510
1.4528	10	310	.8892	8	212
1.4243	2	221	.8774	8	431
1.3598	20	301	.8738	8	332
1.3465	12	112	.8437	6	422
1.3041	2	311	.8292	8	303
1.2441	4	202	.8196	12	521
1.2006	2	212	.8120	2	440
1.1702	6	321	.7877	2	530
1.1483	4	400			
1.1143	2	410			
1.0936	8	222			
1.0827	4	330			

Figure 2.18. The Powder Diffraction File Card No. 21-1276, for rutile, TiO₂. The data on the right-hand side are the *d*-spacings (in Ångströms; 1 Å = 0.1 nm), relative intensities and Miller indices of the reflections in the X-ray powder diffraction pattern. The middle box on the left-hand side gives the crystallographic data. Note that the layout of the computerized version is slightly different. Reproduced with the permission of the International Centre for Diffraction Data (Powder Diffraction Card No. 21-1276).

the unit-cell parameters for the different crystal systems is given in Appendix A.

If you need the atomic positions for the computer program you are using, these can usually be found in the references given for the substance of interest in the Powder Diffraction File. Other sources are: Wells (1984), which contains crystal structure data for inorganic materials, Villars and Calvert (1985), which contains data for inter-metallic phases, and NIST Crystal Data.

2.9.2 *Measuring the pattern and indexing it – an example*

Figure 2.19 is a diffraction pattern from plagioclase feldspar, a framework alumino-silicate. It is triclinic with cell parameters: $a = 0.8169$, $b = 1.2851$, $c = 0.7124$ nm, $\alpha = 93.63$, $\beta = 116.4$, $\gamma = 89.46°$, and the lattice is C-face centred. In order to index the pattern, we need to measure the R values for two spots and convert them to d-values using the equation: $d = \lambda L/R$. We will also need the angle between the two **g**-vectors. Although the program we will use (Windows CM k-Space Control) will cope with any of the R-values in the pattern, it is better to choose the two shortest ones because this will give us fewer possibilities for the solution. It is also important to measure the values as accurately as possible; for instance, by calculating the R-values by measuring across as many spots as possible and measuring where the spots are sharp. This is particularly true for a crystal of low symmetry and large cell parameters such as this where there are a large number of possible d-spacings. The two R-values chosen in *Figure 2.19* and the angle between them are shown. The computer program gives three possible solutions, but the closest match is $[0\bar{1}1]$ *(Table 2.3)*. You will see that the ratio of the d-spacings and the angle for this zone are particularly close matches to the experimental ones. This is because these values are not subject to systematic errors such as an incorrect λL value. A computer generated plot of the $[0\bar{1}1]$ diffraction pattern of plagioclase feldspar is shown in *Figure 2.17*.

Figure 2.19. Diffraction pattern from plagioclase feldspar which is C-face-centred triclinic with cell parameters: $a = 0.8169$, $b = 1.2851$, $c = 0.7124$ nm, $\alpha = 93.63$, $\beta = 116.4$, $\gamma = 89.46°$. The camera constant λL was 3.6 nm mm. As $R_1 = 9.5$ mm and $R_2 = 6.35$ mm, using the equation $\lambda L = dR$, $d_1 = 0.379$ nm and $d_2 = 0.567$ nm.

Table 2.3. Comparison of experimental data and computer match for the diffraction pattern in *Figure 2.19*

Zone	d_1 (nm)	hkl	d_2 (nm)	hkl	d_1/d_2	$g_1\hat{\ }g_2(°)$
	0.379		0.567		0.67	107.0
[0$\overline{1}$1]	0.376	111	0.563	1$\overline{11}$	0.67	106.9
[0$\overline{1}$1]	0.388	1$\overline{11}$	0.586	11$\overline{1}$	0.66	103.7
[3$\overline{1}$4]	0.366	130	0.563	1$\overline{11}$	0.65	103.2

The top line shows the experimental values and the lower three lines show the three zones with the closest match according to the computer.

2.10 Calibration of the camera constant

If you want to measure d-spacings from your diffraction patterns, you will need to know the value of the camera constant λL for the camera lengths you use. Although the camera length, L, will probably be displayed on the console of the microscope when it is operating in the diffraction mode and you can therefore calculate λL, it is only an approximate value (it depends on the lens settings) and it is better to measure the value yourself. To do this, you need to record a diffraction pattern from a specimen with accurately known crystal spacings, such as a thin film of pure polycrystalline gold or aluminium, both of which are face-centred cubic. Specimens of these materials can easily be made by sputtering or direct heating of the metal under vacuum onto a grid covered with a carbon film. The polycrystalline metal gives a ring pattern, as shown in *Figure 2.12*. It is important to ensure that the specimen is accurately focused in the image mode at the eucentric height, that the diffraction astigmatism has been corrected and that the diffraction pattern is focused (Section 1.9). Measurement should be made from the film negative of the diameter (not the radius because it is less accurate) of a strong, sharp ring using a calibrated lens. You will need to know the d-spacing of the line or to work it out from the equation in Section 2.8.1. The d-spacings for the first 10 diffraction rings from Al and Au are shown in *Table 2.4*. Then you use the

Table 2.4. d-spacings (nm) of the first nine diffraction rings from Au and Al; both are face-centred cubic.

Ring number	hkl	Au a = 0.418 nm	Al a = 0.405 nm
1	111	0.234	0.235
2	200	0.202	0.204
3	220	0.143	0.144
4	311	0.122	0.123
5	222	0.117	0.118
6	400	0.101	0.102
7	331	0.093	0.094
8	420	0.090	0.091
9	422	0.083	0.083

relation $\lambda L = dR$, where R is the radius of the ring, to find λL. If you put d into the equation in nm and R in mm, your answer will be in nm mm. Although this is not a standard unit, it is convenient for determining d in nm. Values of a typical calibration of the camera constant are shown in *Table 2.5*.

An accuracy of $\pm 1\%$ in λL should easily be achieved with care, but errors can creep in when the constant is used if the experimental pattern is not recorded under exactly the same conditions. If very high accuracy is required, for instance if you are measuring lattice parameters, the standard material can be evaporated directly onto the specimen as an *in situ* standard. However, the image of the polycrystalline film will be superimposed on the image of the specimen.

2.11 Calibration of the rotation between the image and the diffraction pattern

Unlike the glass lenses used in light microscopes that merely invert the image, magnetic lenses also produce a rotation between the object and the image. (However, in some TEMs this rotation has been removed by the addition of a compensating projector lens, in which case there is a fixed rotation, ideally 0°, between image and diffraction pattern.) As the rotation depends upon the lens excitation, the image rotates on the viewing screen as the magnification is changed. As one of the purposes of using a TEM is to correlate detail in the image with crystallographic information provided by the diffraction pattern, the rotation between the image and the diffraction pattern must be known.

The most common substance that is used to measure the rotation is α-MoO_3. It can be prepared by heating ammonium molybdate to red heat in a crucible. Small, thin crystals are formed on the lid of the crucible and may be washed into an aqueous suspension. A drop of this suspension is allowed to dry on an electron-microscope grid covered with a carbon film.

Table 2.5. Comparison of experimentally measured camera length and camera constant with the digital readout for a Philips CM 200 operating at 200 kV

Camera length, L (mm)		Camera constant,
Nominal	Actual	λL (nm mm)
135	138	0.346
190	194	0.487
265	276	0.692
360	369	0.926
500	491	1.233
700	689	1.730
1000	997	2.502
1350	1357	3.407
1900	1930	4.845

Molybdenum trioxide is pseudo-orthorhombic with lattice parameters $a = 0.3966$, $b = 1.3848$, $c = 0.3696$ nm. The crystals lie on their (010) faces and are elongated along the [001] direction (*Figure 2.20*). The calibration is carried out by recording a double exposure for each magnification, with a diffraction pattern and an image superimposed. The calibration needs to be repeated for each camera length used as there is a rotation of the diffraction pattern with a change in the camera length (see *Figure 1.6*). Examples of values of the rotation angle are given in *Table 2.6*.

Although the above calibration of the rotation angle is adequate for most interpretations of the image in a TEM; for some applications, for instance the determination of the sign of dislocation loops (see for instance Williams and Carter, 1996), it is necessary to allow for any 180° inversion between the image and the diffraction pattern. As explained in Section 1.1 and shown in *Figure 1.1*, a converging lens inverts the image, but not the diffraction pattern, relative to the object. In the electron microscope, whether there is an inversion or not depends upon how many imaging lenses are switched on at a particular magnification; if an odd

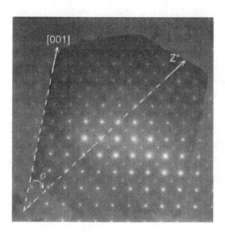

Figure 2.20. A double exposure showing the superposition of a MoO_3 crystal and a [010] diffraction pattern from the same crystal. The rotation, ϕ, between the image and the diffraction pattern is 35.75° clockwise; in other words, the image of the crystal has to be rotated by this angle to bring it into crystallographic coincidence with the diffraction pattern.

Table 2.6. Rotation between the image and the diffraction pattern for a Philips CM 200 operating at 200 kV

Magnification ($\times$ 1000)	$L = 1.9$ m	$L = 1.0$ m
11.5	62°C	49°C
15	56°C	43°C
20	44°C	31°C
27.5	18°C	5°C
38	1°C	12°A
50	6°C	7°A
66	15°C	2°C
88	3°C	10°A

C denotes a clockwise and A denotes an anti-clockwise rotation. These are the angles that the image needs to be rotated to bring it into crystallographic coincidence with the diffraction pattern. Note that these values are not absolute; they vary from instrument to instrument.

number of lenses is excited there will be a resultant inversion, if an even number is excited there will not. One way to see if the image has a 180° inversion is to underfocus the diffraction pattern slightly so that a defocused, low-magnification image can be seen in the diffuse spot that constitutes the undeviated beam. Comparison of this image with that seen in the normal imaging mode will indicate whether an inversion is present or not (*Figure 2.21*).

2.12 Using electron diffraction to identify unknown phases

One of the uses of electron diffraction is to confirm the identity of a phase or to identify an 'unknown' phase. The 'unknown' phase may be new to science or it may be a phase that has been previously described, but was not known to be in the specimen being examined. If the phase is truly new, then its full characterization will require quantitative information about its composition (by quantitative energy-dispersive analysis using the X-ray spectrum, EDX, or by electron energy-loss spectroscopy, EELS; see for instance Williams and Carter, 1996), its cell parameters, its space group (see Chapter 6) and the atomic positions in the unit cell. However, if the phase has been previously documented, its identity can be found from a search of a database of reference compounds, as long as the interplanar spacings of the longest six to 10 *d*-spacings and the angles between them are known (see Lyman and Carr, 1993). The search process is helped considerably if compositional data are available. The database recommended by Lyman and Carr is the NIST/Sandia/ICDD Electron Diffraction Database (EDD) because it is specifically designed for searches on compositional and electron-diffraction data.

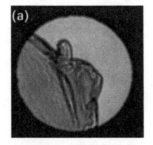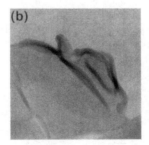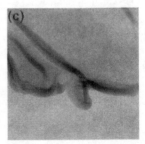

Figure 2.21. The image seen in the underfocused, undeviated beam in a diffraction pattern (a) is compared with the image seen in the normal imaging mode at two magnifications in (b) and (c). The image in the spot is inverted between (a) and (c), but not between (a) and (b).

Exercises

2.1 Assuming that the image formed by a diffracted ray is displaced by $C_s\alpha^3$ (see Section 1.5) in the plane of the object relative to the image formed by the undeviated ray, calculate the displacement for the 222 reflection of aluminium ($a = 0.404$ nm) at 100 kV ($\lambda = 3.7$ pm) in a TEM with a C_s value of 3 mm. Compare this with the displacement for the same reflection in a TEM operating at 300 kV ($\lambda = 1.97$ pm) and with a C_s value of 1 mm.

2.2 *Figure 2.22* shows two diffraction patterns from aluminium, which is F cubic. The camera constant λL is 2.35 nm mm. With the aid of *Figure 2.14c*, identify the zone axis to which the beam is approximately parallel in each case. You will find it helpful to rotate the experimental patterns until you can see their symmetry. When attempting this recognition exercise it is also worth remembering that, as the patterns were both taken at the same camera length, reflections from planes of the same form will have the same R-values. Index the six reflections nearest to the origin in each pattern and calculate the lattice parameter of aluminium.

2.3 Calculate the camera constant, λL, from the powder diffraction pattern in *Figure 2.12a*. Gold has a lattice parameter $a = 0.418$ nm.

2.4 *Figure 2.23* is a [001] diffraction pattern from the mineral olivine, Mg_2SiO_4, which is primitive orthorhombic. The camera constant, λL, is 2.55 nm mm. Calculate the unit cell parameters a and b of olivine.

2.5 *Figure 2.24* is a [010] diffraction pattern from the mineral cummingtonite, approximately $Mg_7Si_8O_{22}(OH)_2$, which is C-face-

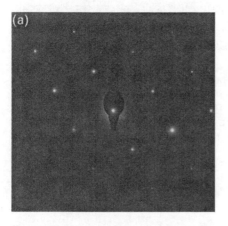

Figure 2.22. Two diffraction patterns from aluminium, which is face-centred cubic.

Figure 2.23. [001] diffraction pattern from the mineral olivine, which is P orthorhombic.

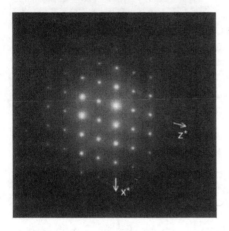

Figure 2.24. [010] diffraction pattern from the mineral cummingtonite, which is C-face-centred monoclinic.

centred monoclinic. The x* and z* axes are marked. Index the eight spots closest to the undeviated beam. Hint: make sure you take into account the systematic absence for this lattice type (*Table 2.1*). Find the lattice parameters a, c and β. The camera constant is 2.38 nm mm. You will find reference to *Figure 2.9* helpful.

2.6 Index the eight reflections nearest to the origin in *Figure 2.19* by reference to the computer-generated pattern in *Figure 2.17*.

2.7 By sketching their respective reciprocal units cells, show that the reciprocal lattice of a crystal with a face-centred direct lattice is body centred, and that the reciprocal lattice of a C-face-centred direct lattice is C-face centred.

3 The reflecting sphere

In order to develop our understanding of electron diffraction further, we need to introduce a very useful concept that links the reciprocal lattice to Bragg's Law. This is the **reflecting sphere**, otherwise known as the **Ewald sphere** because it was first described by P.P. Ewald in 1919 as an aid to the interpretation of diffraction patterns.

3.1 The reflecting sphere

The sphere has a radius of $1/\lambda$, where λ is the wavelength of the electrons, it passes through the origin of the reciprocal lattice, 000, and one of its diameters is the direction of the direct beam (*Figure 3.1*). If a reciprocal lattice point, P, lies on the sphere, we will see that it represents a set of planes that are at the correct angle θ to the direct beam for diffraction to occur, i.e. Bragg's Law is obeyed.

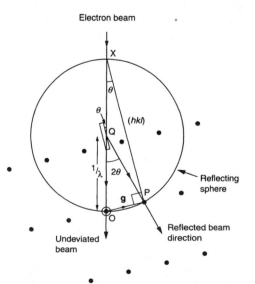

Figure 3.1. The reflecting sphere construction. The sphere has a radius of $1/\lambda$ and the diameter, XO, is the direct beam passing through the origin of the reciprocal lattice, O. If a reciprocal lattice point P lies on the surface of the sphere, Bragg's Law is satisfied, i.e . the set of planes represented by P are diffracting. The direction of the diffracted beam is QP. Adapted from Steadman R, *Crystallography*, 1982, with the permission of van Nostrand Reinhold, New York.

In *Figure 3.1*, <u>OP</u> is the reciprocal lattice vector, $\mathbf{g}_{hkl}$, and OP = $|\mathbf{g}| = 1/d$. Therefore, as the angle XPO is a right angle, XP must represent the trace of the plane (hkl), and it follows that OXP is equal to θ, the angle between the diffracting plane and the incident beam. From triangle OXP:

$$\sin \theta = OP/OX = \frac{1/d}{2/\lambda}$$

or, rearranging:

$$2d \sin \theta = \lambda$$

which is Bragg's Law. Thus, whenever a reciprocal lattice point touches the sphere a diffracted beam appears.

The reflecting sphere construction has another property that is useful for visualizing the diffraction process. As the angle OQP is equal to 2θ (the angle at the centre is twice the angle at the circumference), QP is the direction of the diffracted ray and Q can be considered as the position of the crystal in real space. We can simplify the construction of the reflecting sphere by showing only the points Q, O and P, as shown in *Figure 3.2*. This diagram also allow us to write a vector equation as an alternative to Bragg's Law. If the direct beam is a vector $\mathbf{k}_0$ of modulus $1/\lambda$ and the diffracted beam is a vector $\mathbf{k}$, also with a modulus $1/\lambda$, we can write:

$$\mathbf{g} = \mathbf{k} - \mathbf{k}_0$$

If the reciprocal lattice point has to be exactly on the reflecting sphere for diffraction to occur, you may be wondering how it is that, when the electron beam is parallel to a zone axis, we see on the screen of the TEM a diffraction pattern that consists of a large number of reflections. There are two factors that explain this paradox. Firstly, the radius of the reflecting sphere is extremely large in electron diffraction in comparison with the size of the reciprocal lattice. This is a consequence of the very small wavelength of the electrons. The geometry of *Figure 3.2* is more

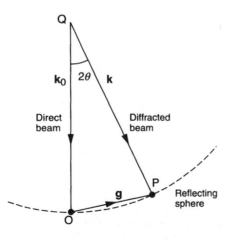

Figure 3.2. The reflecting sphere construction and Bragg's Law in vector notation. $\mathbf{k}_0$ is the direct beam and $\mathbf{k}$ is the diffracted beam. If the reciprocal lattice point P is on the sphere, then $\mathbf{g} = \mathbf{k} - \mathbf{k}_0$.

appropriate to X-ray diffraction, where the wavelength is of the same order of magnitude as atomic dimensions, than to electron diffraction, where the wavelength is two orders of magnitude smaller than this (Appendix D). A more realistic diagram for the geometry of electron diffraction is shown in *Figure 3.3.* As you can see, a large number of reciprocal lattice points are close to the surface of the reflecting sphere because the radius of the sphere is so large.

The second main reason that we see a diffraction pattern on the screen of the electron microscope without having to move the specimen with respect to the sphere is that the diffraction 'spots' are not really spots, but are elongated perpendicular to the surface of the specimen. This is a result of the fact that the specimen has to be very thin, $\leqslant \sim 0.2$ μm, for it to be able to transmit electrons, and for an image or diffraction pattern to be recorded on the screen of a TEM. If the specimen is of thickness t, the reciprocal lattice point is streaked out to a length of $1/t$ (see Chapter 7), as shown in *Figure 3.4a,* producing a **relrod** (reciprocal lattice rod). Now we can see that many relrods intersect the sphere if the electron beam is travelling along a zone axis, and a lattice of spots results on the screen of the TEM. If the reciprocal dimension parallel to the beam is fairly small (the real-lattice repeat is fairly large), we may even see spots from the reciprocal lattice layer above, the so-called **first-order Laue zone,** as an annulus of spots about the zero layer (the plane of reciprocal lattice points that passes through the origin 000 is known as the **zero-order Laue zone**). Even if they are not visible at the exact zone-axis, the higher-order Laue zones will become evident if we tilt the specimen by a small angle away from the beam being exactly parallel to the zone axis (*Figures 3.4b* and *3.5*). In such a pattern, the origin of the reciprocal lattice (the undiffracted beam) is no longer at the centre of the array of spots. You will also notice that, as the crystal is tilted, the intensities of the reflections change (*Figure 3.5*). This is because the intensity along the diffraction streak varies as shown in *Figure 3.6.* The intensity is at a maximum at the centre of the streak, i.e. when the 'spot' is exactly on the reflecting sphere. In *Figure 3.5b*, it is easy to see where the reflecting

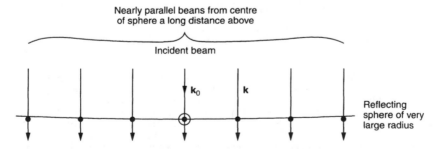

Figure 3.3. A section of the reflecting sphere and the reciprocal lattice at approximately their correct relative sizes for the diffraction of electrons diffracted at 200 kV. A large number of reciprocal lattice points are close to the surface of the sphere because its radius is so large.

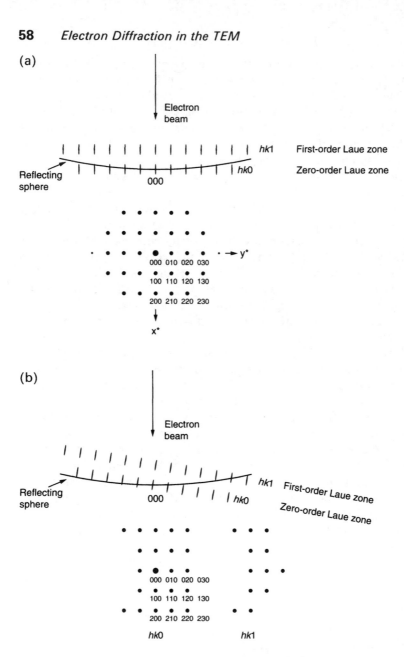

Figure 3.4. (a) Diagram of the reciprocal lattice and reflecting sphere in which the reciprocal lattice 'points' are shown elongated perpendicular to the thin dimension of the crystal by a distance $1/t$, where t is the thickness. The upper diagram shows the electron beam along [001], i.e. perpendicular to the $hk0$ and $hk1$ layers of the reciprocal lattice. The lower diagram shows the resultant [001] diffraction pattern of an orthorhombic crystal. (b) The formation of an electron diffraction pattern, as in (a), but with the crystal tilted slightly away from the exact zone axis. The reflecting sphere passes through a smaller area of the zero-order Laue zone, the $hk0$ layer, than before, but now passes through part of the first-order Laue zone, the $hk1$ layer. The resulting diffraction pattern is shown below. Reproduced from Putnis A, *Introduction to Mineral Sciences*, 1992, with the permission of Cambridge University Press, Cambridge.

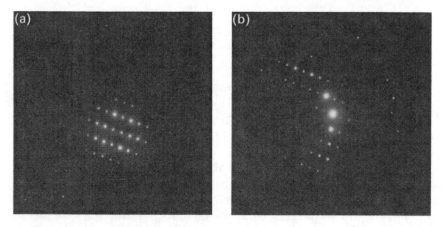

Figure 3.5. Experimental electron diffraction patterns. (a) The electron beam is parallel to the zone axis. (b) The crystal has been tilted slightly with the consequence that the first-order Laue zone is visible.

sphere passes through the exact Bragg position from the ring of bright spots in the zero layer.

Figure 3.6 allows us to introduce the vector parameter **s**, which is a measure of the deviation of the reciprocal lattice point from the exact Bragg position. **s** is known as the **deviation parameter** or the **excitation error**. When the diffraction spot is not exactly at the Bragg condition we can write in vector notation:

$$\mathbf{g} + \mathbf{s} = \mathbf{k} - \mathbf{k}_0.$$

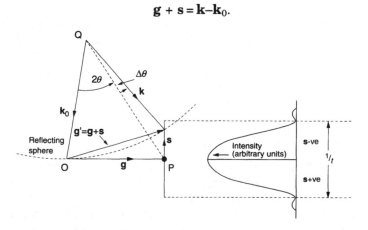

Figure 3.6. Geometry of electron diffraction when a reflection deviates from the exact Bragg condition by an angle $\Delta\theta$. The diagram to the right shows the intensity of the diffracted beam as a function of **s**, the deviation vector of the reflection from the reflecting sphere, for a constant thickness of the crystal, *t*. **s** is positive when the spot is inside the sphere and negative when it is outside the sphere. The effective length of the streak is $1/t$ where *t* is the thickness of the crystal.

3.2 Obtaining orientation relationships

One of the most useful features of the electron microscope is the ability to form a diffraction pattern from finely intergrown phases, planar defects or dislocations (a dislocation is a line defect that distorts the lattice around it), and to be able to determine their crystallographic relationship; in other words, to describe which planes and directions are parallel in the two phases, on which planes, called **habit planes** (see Section 3.3), the defects lie or in which directions the dislocations lie. Electron diffraction, together with the associated image, is also used to determine the fault vectors of dislocations (burgers vectors) and stacking faults. These latter applications are beyond the scope of this book; details can be found in Williams and Carter (1996).

3.2.1 Second-phase particles or precipitates

Orientation relationships between phases are usually described in terms of a pair of parallel directions in a pair of parallel planes. The determination of such an orientation relationship is best illustrated with an example. *Figure 3.7a* is a diffraction pattern from as-cast nickel–aluminium bronze (Cu with approximately 10 wt. % Al, 5 wt. % Ni, 5 wt. % Fe) and Ni-rich κ_{III} precipitates. The matrix is face-centred cubic and the precipitate (the fainter spots) is primitive cubic. It is clear from a casual inspection of the pattern that there is an orientation relationship between the two phases and that, because of the coincidence of many of the spots, the lattice parameters are related in a rational way. The two patterns are indexed in *Figure 3.7b*. Both patterns are from $<\bar{1}12>$ zone axes (compare with *Figure 2.14a* and *c*). You will also notice that the **g**-vectors for the (220) planes of the matrix and the $(1\bar{1}1)$ planes of the precipitate are parallel. Therefore, these two sets of planes must be parallel in the real lattice. The orientation relationship can therefore be written as:

$$(110)_{\text{matrix}}//(1\bar{1}1)_{\text{precipitate}}; \ [\bar{1}12]_{\text{matrix}}//[\bar{1}12]_{\text{precipitate}}$$

(remember that the (110) planes are parallel to the (220) planes). You will also notice that the $\bar{1}1\bar{1}$ spot from the matrix and the 110 spot from the precipitate coincide. Thus, we could also have written: $(\bar{1}1\bar{1})_{\text{matrix}}//$ $(110)_{\text{precipitate}}$. I should emphasize that the above relationship is only one way of expressing the orientation relationship between the phases. The zone axes, for instance, could be described by the symbols of any one of the 24 zones in the set $<112>$. It is conventional to give one specific orientation relationship rather than expressing the planes as forms, $\{hkl\}$, and the zones as belonging to one of a set $<UVW>$. The indexing of the planes and directions should be self-consistent, i.e. the quoted direction should lie in the quoted plane, as above.

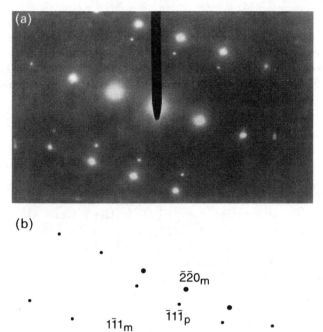

Figure 3.7. (a) Diffraction pattern from an as-cast nickel–aluminium bronze alloy. The matrix is face-centred cubic, $a = 0.36$ nm and the precipitate is primitive cubic, $a = 0.29$ nm. (b) The indexed pattern. The matrix reflections are shown as large dots and the precipitate spots are shown as smaller dots. Diffraction pattern reproduced from Hasan F *et al.*, *Metall. Trans.*, 1982; 13A: 1337–1345, with permission of the Minerals, Metals and Materials Society/ASM International.

From the coincidence of the $\bar{1}1\bar{1}_{\text{matrix}}$ and $110_{\text{precipitate}}$ spots we can conclude that the d-spacings for the corresponding planes are equal, i.e. $d_{111m} = d_{110p}$, and, using the equation relating d to a, h, k and l (Appendix C), that $a_{\text{matrix}}/a_{\text{precipitate}} = \sqrt{3}/\sqrt{2} = 1.22$. (The cell parameters of the two phases are 0.36 and 0.29 nm.)

3.2.2 Twinned crystals

A **twinned crystal** is one which consists of two or more parts that are oriented in a specific crystallographic relationship to each other. The relationship is normally one in which there is a reflection across a plane (the **twin plane**) or a rotation of 180° about the **twin axis** between the lattices of the two individuals. *Figure 3.8* shows a diagram of such a twinned crystal in which the relationship is a reflection in the (110) plane. It can also be described as a 180° rotation about the normal to the (110) plane, though this is less obviously apparent.

Because a twinned crystal consists of two (or more) orientations of the real lattice, the diffraction pattern from it will show one (or more) superimposed single-crystal patterns. A diffraction pattern in which the twin plane is normal to the plane of the pattern can be used to identify the twin plane. *Figure 3.9a* is a diffraction pattern from a twinned region of a Ti–6Al–4V alloy which has the hexagonal close-packed structure. *Figure 3.9b* identifies the reciprocal lattices from the two twin individuals. Comparison of one of the patterns with *Figure 2.15e* shows that the zone axis is [121] and that the second pattern can be derived from the first by a reflection in the plane normal to the **g**-vector for the ($\bar{1}011$) planes. Therefore, the twin plane is ($\bar{1}011$). There are 12 planes in the form {$\bar{1}011$} in the crystal class 6/*mmm* – ($\bar{1}011$), ($10\bar{1}1$), ($10\bar{1}\bar{1}$), ($\bar{1}01\bar{1}$), ($1\bar{1}01$), ($\bar{1}10\bar{1}$), ($\bar{1}101$), ($1\bar{1}0\bar{1}$), ($0\bar{1}11$), ($0\bar{1}1\bar{1}$), ($0\bar{1}11$), ($01\bar{1}\bar{1}$), i.e. 12 planes that are equivalent in terms of symmetry. Thus, we would expect there to be *six* orientations of the twins: six rather than 12 because the ($\bar{1}011$) and ($10\bar{1}\bar{1}$) planes etc. are parallel.

3.3 Determining habit and composition planes

The physical interface between intergrown phases such as precipitates or stacking faults is known as the habit plane and the interface between twin individuals is known as the **composition plane**. Determination of these interfacial planes is an important element in the characterization of the material containing them. The easiest way to determine the orientation of a planar interface such as the composition planes in *Figure 3.9c* is to tilt the specimen so that the interface is vertical, i.e.

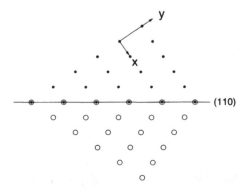

Figure 3.8. Lattice of a twinned orthorhombic crystal. The twin plane is (110).

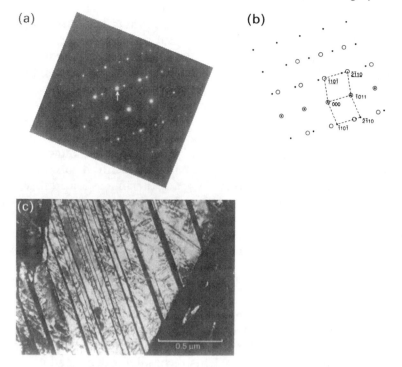

Figure 3.9. Twinning in a Ti–6Al–4V alloy which has the hexagonal close-packed structure, $a = 0.2951$, $c = 0.4684$. (a) Diffraction pattern from the twins. (b) Diagram that identifies the spots from the two twin individuals. One reciprocal lattice is shown as dots and the other as open circles. Underlining of Miller indices is a standard way of indicating that the reflection comes from a twin. The zone axis is [121] = [$0\overline{1}11$] and the twin plane is identified as ($\overline{1}011$). (c) Image of the twins in which their composition plane is parallel to the electron beam. The image is in the correct relative orientation to the diffraction pattern and shows that the composition plane of the twins is also ($\overline{1}011$). Note the streaking in the diffraction pattern normal to the composition planes resulting from their narrow width (see Section 7.2). The image was recorded in the dark-field mode using a spot from one of the twinned individuals (arrowed). As a consequence only one set of twins appears bright in the micrograph. Diffraction pattern courtesy of G.W. Lorimer.

parallel to the electron beam. This is best done in a tilt–rotate stage that has a large range of tilt, usually $\pm 60°$. The procedure is to rotate the sample in the holder until the trace that the interface makes with the sample surface is parallel to the tilt axis. In a side-entry stage, the tilt axis is parallel to a traverse direction of the specimen (the left-hand control in the FEI/Philips range of microscopes). You need to rotate the specimen until a slight movement of the specimen transverse that is parallel to the tilt axis moves the image of the interface along its length. Then, having made the specimen eucentric, the specimen should be tilted until the image of the precipitate or defect shows minimum thickness and the interface appears sharp. The interface should then be vertical. It is now a simple matter to determine the Miller indices of the interface from the diffraction pattern; taking into account the rotation between the image and the diffraction pattern (see Section 2.11), the interface will be perpendicular to the **g**-vector from the planes forming the interface. The twin interfaces in *Figure 3.9c* are vertical and the micrograph has been rotated so that it is in the correct relative orientation to the diffraction pattern in *Figure 3.9a*. The **g**-vector for the (10$\overline{1}$1) planes is perpendicular to the composition planes so we can conclude that the composition plane of the twins is {10$\overline{1}$1}. The twin and composition planes are the same in this case. This is usually, but not always, the case.

A further example of the determination of a habit plane in this way is shown in *Figure 3.10*, which shows small, Cu-rich platelets, known as G-P zones, in a Al–4.5% Cu alloy. The matrix Al is face-centred cubic and the precipitates have the same structure. Again, the image and diffraction pattern are in their correct respective orientation and the precipitate interfaces are vertical. The diffraction pattern is from the [001] zone axis and the two sets of precipitates have interfaces parallel to (100) and (010). Because the (001) planes are symmetrically equivalent to (100) and (010) (the form {100}) there are precipitates parallel to (001) as well. These are not easily seen in *Figure 3.10b* because they are perpendicular to the electron beam.

In cases where the above technique is not feasible, for instance if it is not possible to tilt the interfaces to the vertical or if the defect is contained within the thin specimen and does not intersect the surface, then **trace analysis** is appropriate. This technique is described in detail by Edington (1975) and Hammond (1993).

3.3.1 The importance of crystal orientation

The diffraction patterns that I have used to illustrate the determination of orientation relationships and habit planes are from low-index (high-

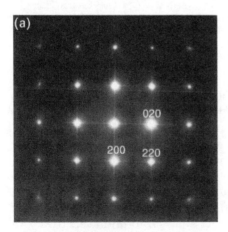
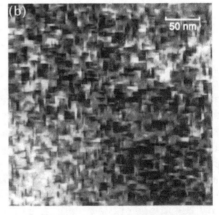

Figure 3.10. Platelets of Cu-rich G-P zones in a Al–4.5% Cu alloy. The Al is face-centred cubic and the precipitates have the same structure as the matrix. (a) [001] diffraction pattern. No separate diffraction spots are seen from the precipitates, but their small width has resulted in streaks perpendicular to their interfaces (see Section 7.2). (b) The image is in correct relative orientation to the diffraction pattern. Diffraction pattern courtesy of P.P. Prangnell.

symmetry) zones. These zones are normally the most informative and it is usually only necessary to find one such zone axis in order to be able to determine the orientation relationship unambiguously. Procedures for tilting the sample so that the electron beam is parallel to a particular zone axis are described in Section 4.3. However, there may be times when you can find a parallel zone axis for each phase, but there are no common **g**-vectors indicating planes that are parallel in each phase. In this case, it will be necessary to tilt to a second zone axis and to record the tilt angle between the two. Any planes that are parallel in the two phases can then be determined by plotting the information on a **stereographic projection** as described by Williams and Carter (1996) and Hammond (1993).

3.3.2 *Dark-field imaging*

The normal method of forming an image in the TEM is to insert the objective aperture around the incident beam in the back focal-plane of the objective lens and to project the image of the specimen formed by the lens onto the viewing screen (*Figure 1.4a*). This procedure excludes most, if not all, of the diffracted beams from interfering in the image plane and areas that are diffracting strongly appear darker in the image than those that are not. In particular, areas where there is no specimen appear bright; hence, the method is called **bright-field imaging**.

If a diffraction pattern consists of the superimposed patterns from more than one phase, the technique of **dark-field imaging** may be used to determine which spots originate from which phases. The technique consists of allowing a single diffracted beam to pass through the objective aperture, such that the bright areas in the image come from the phase responsible for the spot. The image of the twins in *Figure 3.9* was taken in the dark-field mode using a spot from one of the twin individuals (the one that is arrowed in *Figure 3.9a*). As a consequence, only one set of twins appears bright in the image.

Although a dark-field image can be formed by displacement of the objective aperture from the optic axis to the diffraction spot (*Figure 3.11a*), the image is of poor quality because aberrations and astigmatism are introduced. However, most modern microscopes allow the incident beam to be tilted so that the diffracted beam passes along the optic axis (*Figure 3.11b*). This is called **centred dark-field imaging**. The method is as follows.

- Adjust the specimen height so that it is eucentric.
- Obtain a bright-field image and focus it.
- Obtain a diffraction pattern from the desired area.
- Decide which *hkl* reflection you wish to use for the dark-field image. Tilt the specimen so that the spot is as bright as possible. The (*hkl*) planes should then be at the Bragg angle θ to the incident beam.
- Underfocus the second condenser lens and mark the position of the incident beam with the beam stop.

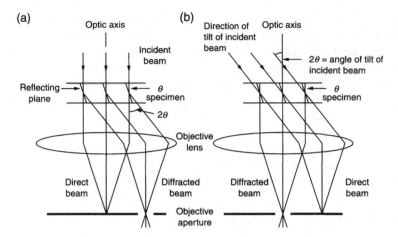

Figure 3.11. The formation of dark-field images. (a) The objective aperture is displaced so that it allows an off-axis diffracted beam to pass through it. This produces a poor-quality image. (b) The incident beam is tilted so that the diffracted beam is parallel to the optic axis. This produces an image with high resolution. Adapted from Williams DB and Carter CB, *Transmission Electron Microscopy*, 1996, with the permission of Plenum Press, New York.

- Switch to the dark-field mode and use the dark-field tilt controls to bring the diffraction spot diametrically opposite to the chosen one onto the position on the screen where the incident beam was previously positioned. This spot will now be strong and the $(\overline{hkl})$ planes will be at θ to the incident beam. The hkl reflection will be weak.
- Switch off the dark-field deflectors and insert and centre a small objective aperture.
- Switch the dark-field deflectors back on and adjust the dark-field tilt controls so that the $\overline{hkl}$ reflection is in the centre of the aperture. This should be the only reflection visible in the aperture.
- Switch to the image mode. The image is best focused in the bright-field mode.

If you find that the illumination moves during this procedure, it is likely that the tilt coils are misaligned. They may be realigned according to the manufacturer's manual.

Exercises

3.1 *Figure 3.12* is a diffraction pattern from an α-iron specimen which is body-centred cubic, $a = 0.2866$ nm, containing a face-centred cubic precipitate of Fe_2TiSi. Index the patterns from the two phases by comparison with *Figure 2.14b* and *c*, and determine the orientation relationship between them. Calculate the lattice parameter of the precipitate.

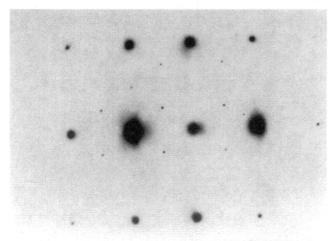

Figure 3.12. Diffraction pattern from an α-iron specimen which is body-centred cubic, $a = 0.2866$ nm, containing a face-centred cubic precipitate of Fe_2TiSi. Reproduced from Hammond C, *The Basics of Crystallography and Diffraction*, 1997, with the permission of Oxford University Press.

3.2 *Figure 3.13* is a representation of the diffraction pattern from as-cast nickel–aluminium bronze (Cu with approximately 10 wt. % Al, 5 wt. % Ni, 5 wt. % Fe). The matrix is face-centred cubic, $a = 0.36$ nm and the Fe-rich precipitate, κ_{IV}, is primitive cubic, $a = 0.29$ nm. The matrix spots are shown as dots and the precipitate spots as open circles. Index both patterns with the help of *Figure 2.14a* and *c*, and determine the orientation relationship between the matrix and the precipitate.

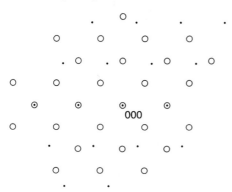

Figure 3.13. Representation of a diffraction pattern from as-cast nickel–aluminium bronze. The matrix is face-centred cubic and the precipitate is primitive cubic. The matrix spots are shown as dots and the precipitate spots as open circles.

4 Finding your way around reciprocal space: Kikuchi diffraction

All the diffraction phenomena that we have looked at so far have been the result of elastically scattered electrons, i.e. those that have not lost energy. However, as the thickness of the specimen increases, some of the electrons will lose a small amount of energy ($\leqslant 50$ eV). The effects of these **inelastically scattered electrons** are visible in most spot patterns, such as those shown in *Figures 2.24* and *3.5*, both as a diffuse halo around the undeviated beam and as an overall faint background intensity. The inelastically scattered electrons can subsequently be elastically scattered (i.e. Bragg diffracted) by lattice planes and produce a phenomenon known as **Kikuchi lines**. Kikuchi lines will be best seen in diffraction patterns from areas of the specimen that have a low density of defects and are about half the thickness that the beam can penetrate or thicker. If the specimen is thinner than this, only spots will be seen; if it is very thick, only Kikuchi lines will be seen (see *Figures 4.5* and *4.8*), whereas areas of intermediate thickness will show both spots and Kikuchi lines (*Figure 4.1*). Kikuchi patterns are very useful to the electron microscopist because, as we shall see, they facilitate finding your way around reciprocal space and determining the exact orientation of the crystal.

A full treatment of the origin of Kikuchi lines requires use of the theory of Bloch waves, but for our purposes the much simplified explanation given below will suffice.

4.1 How Kikuchi lines form

The intensity distribution of the inelastic spectrum as a function of scattering angle is shown in *Figure 4.2*; it is at a maximum in the forward direction and falls off with scattering angle. In *Figure 4.3a* a set of lattice planes (hkl) are shown close to, but not exactly at, the Bragg angle θ to

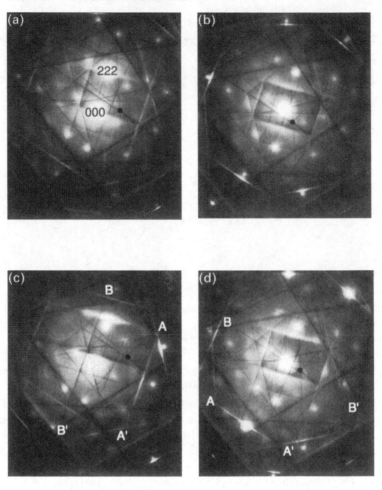

Figure 4.1. Diffraction patterns from austenitic stainless steel (face-centred cubic) showing spots and Kikuchi lines. Note the pairs of parallel light and dark lines such as AA′ and BB′. The incident beam is approximately parallel to [1$\bar{1}$0] and the strongly diffracting reflection is 222. There is a small tilt between each pattern; note how this causes a particular crossover point on the pattern (marked with a dot) to move around the pattern. (a) 222 is exactly on the reflecting sphere ($s = 0$). (b) The 222 and $\overline{222}$ reflections are equidistant from the sphere (the symmetry position). (c) The excess Kikuchi line is on the far side of the 222 spot from 000, so s is positive. (d) The excess Kikuchi line is on the near side of the 222 spot from 000, so s is negative.

the incident beam. However, two rays from the inelastic spectrum (see *Figure 4.2*) are incident on the planes at the Bragg angle and will be diffracted by them as shown. As ray 1 is closer to the forward direction than ray 2, it is more intense (*Figure 4.2*), and an excess number of electrons over the background will arrive in the back focal-plane at B; correspondingly, there will be a deficiency of electrons at D. Thus, there is a bright line at B and a dark line at D in the diffraction pattern; these are the Kikuchi lines. The bright line is always the one further from the

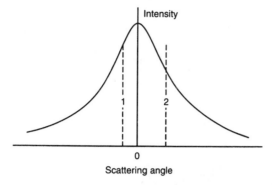

Figure 4.2. The intensity of the inelastic scattering as a function of scattering angle.

incident beam direction 000. Strictly, if ray 1 is incident on the (hkl) planes, then ray 2 is incident on the $(\overline{hkl})$ planes, as the two sets of indices refer to the same sets of planes, but viewed in the opposite sense. Thus, in *Figure 4.3a* the bright Kikuchi line is associated with the hkl or $+\mathbf{g}$ reflection and the dark Kikuchi line is associated with the $\overline{hkl}$ or $-\mathbf{g}$ reflection.

The diffracted rays actually form cones of semi-angle $90-\theta$, called **Kossel cones**, and what we see in the diffraction pattern is a pair of parabolas where the cones intersect the Ewald sphere. The parabolas appear as straight lines in the diffraction pattern because the angles involved are very small. The $\pm\mathbf{g}$ pair of lines and the region between them is known as a **Kikuchi band**.

The angular separation of the pair of lines is 2θ, their spatial separation in the diffraction pattern in the back focal-plane is g, which of course is the same as the distance of the diffraction spot from the origin of the pattern at 000, and they are perpendicular to the **g**-vector. It follows that, because a Kikuchi line is parallel to the plane with which it is associated, the angle between any two lines is equal to the angle between the corresponding planes. Each reflection has an associated pair of Kikuchi lines and, if the specimen is tilted, they move as if rigidly attached to it. If the specimen is tilted by a small angle α, the lines will move a distance r across the pattern on the screen, and, from simple geometry, $r=L\alpha$, where L is the camera length. For a camera length of 1000 mm, this means a shift of about 30 mm per degree of tilt. The change in the spot pattern for such a small angle of tilt is nothing like as dramatic as this, as can be seen in *Figures 4.1* and *3.5*; the intensities of the spots change as they move closer to or further away from the reflecting sphere, but their positions do not. It is the dramatic movement of the Kikuchi lines with a small angle of tilt that allows us to use them to determine the exact orientation of the crystal to within about 0.1 degree, as shown below.

In order to characterize certain defects such as dislocations and stacking faults, it is important to set up **two-beam conditions** in which only one set of planes is strongly diffracting, i.e. the relevant reciprocal lattice point lies exactly on the reflecting sphere and the deviation parameter $s=0$. It is also important that s is positive if you want to form images of defects with good contrast (see Williams and Carter, 1996).

(a)

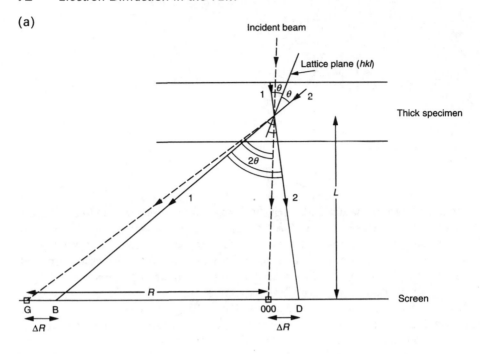

(b)

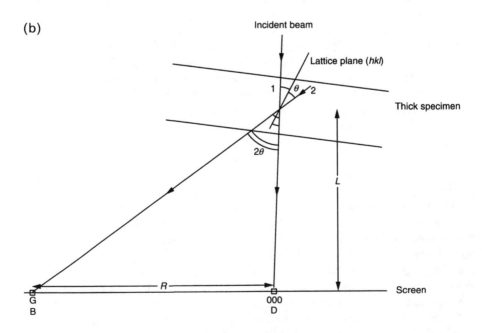

Figure 4.3. Ray diagram showing the geometry of Kikuchi lines. (a) The lattice plane (*hkl*) is close to, but not exactly at, the Bragg angle *θ* to the incident beam. G is the diffraction spot *hkl*. (b) A clockwise rotation of the crystal from its position in (a) has brought the lattice plane to the Bragg condition with respect to the incident beam. Now the bright Kikuchi line passes through the diffraction spot G and the dark line passes through the origin at 000.

Kikuchi lines enable you to carry out these tasks, to find the exact orientation of the crystal or of orientation changes, such as across a low-angle grain boundary, and to orient the crystal to take a pair of stereo images (see Williams and Carter, 1996); none of these things are possible with the spot pattern alone.

Figure 4.4 is a schematic that shows, on the left, four different positions of the diffraction spot relative to the reflecting sphere and, on the right, the positions of the corresponding Kikuchi lines in the diffraction pattern. If you imagine the specimen in *Figure 4.3a* being tilted clockwise until the bright Kikuchi line 1 passes through the diffraction spot G, then the dark Kikuchi line 2 will pass through the central spot 000, as shown in *Figure 4.3b*, and the lattice plane (*hkl*) will be at the exact Bragg angle θ to the incident beam, the deviation parameter $s = 0$ and the diffraction spot will be at its brightest. Thus, to produce two-beam conditions, it is simply a matter of moving to an area of the sample that is thick enough to show both spots and Kikuchi lines in the diffraction pattern, and tilting the specimen until the bright Kikuchi line associated with the spot passes through it and the dark line passes through the centre spot. This is the situation for the 222 spot in *Figure 4.1a*.

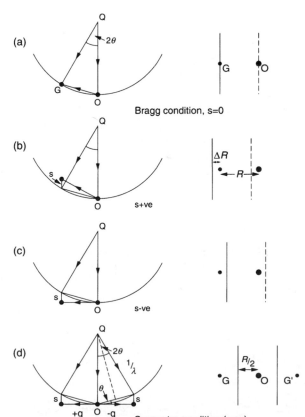

Figure 4.4. The positions of Kikuchi lines in the diffraction pattern (on the right) and the corresponding position of the diffraction spot relative to the reflecting sphere (on the left). The full line represents a bright Kikuchi line and the dashed line represents a dark Kikuchi line. At the symmetry position in (d) the lines are of equal intensity.

In *Figure 4.4b*, the diffraction spot is inside the reflecting sphere, so s is positive. This situation can be arrived at from that in *Figure 4.4a* by a clockwise rotation of the crystal at Q. A clockwise rotation of the crystal in *Figure 4.3a* from the exact Bragg condition will move the Kikuchi lines so that the dark line lies between 000 and G and the bright line lies on the far side of G from 000, as shown on the right-hand side of *Figure 4.4b*. In *Figure 4.1b*, s is positive for the 222 reflection. Conversely, if s is negative (spot outside the reflecting sphere) the Kikuchi lines are displaced so that the dark Kikuchi line is on the far side of 000 from G, as shown in *Figures 4.4c* and *4.1d*.

When the hkl and $\overline{hkl}$ reflections are at an equal distance from the reflecting sphere, as in *Figure 4.4d*, we say that we have the **symmetry condition** for those reflections. Our simple explanation of the formation of Kikuchi lines breaks down here because it predicts that, because they are equidistant from G or G′ and O, the two Kikuchi lines should be of equal intensity and indistinguishable from the background. That this is not the case is clear from *Figure 4.1b*; there is a band of higher intensity bounded by the Kikuchi lines.

4.2 The determination of the deviation parameter s

The deviation parameter s is important in the quantitative analysis of the contrast seen in images of crystalline material. As we have seen, the sign of s can be obtained from inspection of the diffraction pattern if it contains Kikuchi lines. In addition, the magnitude of s can be obtained from the displacement, ΔR, of the Kikuchi line from the corresponding diffraction spot (*Figure 4.3a*). At the symmetry position, if θ is expressed in radians and because θ is small, we can write (*Figure 4.4d*):

$$s \sim g\ \theta.$$

Also: $\qquad\qquad\qquad\quad g/2 \sim 1/\lambda \times \theta.$

Rearranging: $\qquad\qquad\quad s = g^2 \lambda / 2.$

At the symmetry position, $\Delta R = R/2$ and at the exact Bragg condition, $s = 0$ and $\Delta R = 0$. Therefore, by simple proportions:

$$s/\Delta R = g^2 \lambda / 2 \div R/2$$

or: $\qquad\qquad\qquad\quad s = g^2 \lambda \times \Delta R / R.$

4.3 Kikuchi maps and their uses

We have seen that every pair of planes (hkl) and $(\overline{hkl})$ gives rise to a Kikuchi band, i.e. a pair of bright/dark lines, and that, when the planes are parallel to the incident beam, the band is symmetrical about the spot from the undeviated beam. It follows that, at the exact zone axis, the Kikuchi bands from *all* the planes in the zone are centred on the undeviated beam (*Figure 4.5*). Thus, if we wish to orient the crystal so that the electron beam is accurately parallel to the zone axis, we need to tilt the crystal so that all the Kikuchi bands intersect at the centre of the pattern. When the electron beam is close to a zone axis, as in *Figure 4.6a*, the position of the zone axis relative to the undeviated beam can readily be identified. Note that it is easier to see Kikuchi lines if the diffraction aperture is removed and the electron beam is focused more than normal (see Williams and Carter, 1996, Fig. 20.11). This is because a convergent beam samples a smaller area than SAD and a small area is less likely to be deformed.

Kikuchi patterns taken exactly along a zone axis have the useful property that they indicate the symmetry of the crystal along that axis. This is illustrated by *Figure 4.5*, which compares the symmetry shown in the Kikuchi patterns by the [111] germanium and [001] = [0001] magnesium zone axes. Ge is cubic with the diamond structure and Mg is close-packed hexagonal. Both spot patterns show hexagonal symmetry (see *Figures 2.14* and *2.15*), but the [111] Kikuchi pattern from Ge shows three-fold symmetry, whereas the [001] pattern from Mg shows six-fold symmetry.

 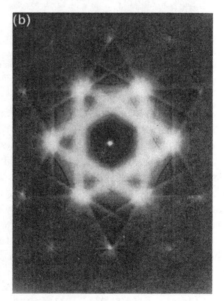

Figure 4.5. Kikuchi patterns in which the electron beam is exactly along a zone axis. In both cases, the specimen is too thick to show diffraction spots. (a) The [111] zone of Ge, showing three-fold symmetry. The narrowest Kikuchi bands correspond to the {220} planes. This pattern can be compared with the schematic Kikuchi map in *Figure 4.7*. (b) The [001] zone of Mg, showing six-fold symmetry. The prominent bands originate from the {10$\bar{1}$0} planes. Reproduced from Levine E *et al.*, *J. Appl. Physiol.*, 1966; 37: 2141–2148, with permission of the American Institute of Physics.

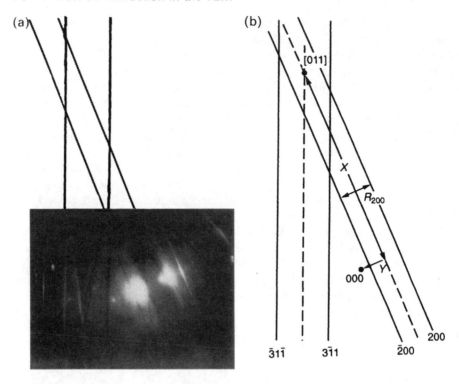

Figure 4.6. (a) A diffraction pattern taken at 100 kV from aluminium with the beam direction close to [011]. (b) Schematic diagram showing the 200 and 3̄11 Kikuchi bands and the position of the undeviated beam 000. The zone direction is at a tilt of 7.5° about the normal to the (200) planes together with a tilt of 0.98° about the normal to the (022̄) planes.

Kikuchi maps are schematic representations of Kikuchi patterns for a number of zones in which the Kikuchi bands are drawn with a width proportional to g. Maps are published in the literature for the primitive, face-centred and body-centred cubic lattice types, the diamond-cubic structure, and for the hexagonal close-packed structure (e.g. *Figure 4.7*). They may be computed for any other crystal lattice type using some of the computer packages listed in Appendix F, or you can compile your own experimentally by photographing the Kikuchi pattern at each prominent zone. These maps can be used to recognize zone axes from their symmetry, at least in the simpler crystal systems. They can also be used rather like road maps to tilt the specimen to a desired zone. For instance, if you recognised from its Kikuchi or spot pattern that a diamond-cubic crystal was oriented with the [233] zone axis approximately parallel to the incident electron beam, you would know from the map in *Figure 4.7* that tilting the sample along the $\pm 02\bar{2}$ Kikuchi band (the narrowest one) would bring you to the [011] or [111] zone, depending on the direction of tilt. In fact, tilting along the narrowest Kikuchi band (which represents the set of reflections with the largest d-value in the zone) will almost always bring you to a more prominent, lower index, zone (unless you are

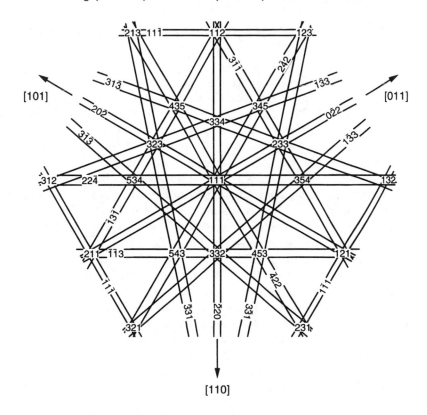

Figure 4.7. A schematic Kikuchi map for diamond-cubic crystals showing zone axes within about 20° of [111]. The indices of some of the prominent Kikuchi bands and zones are shown. Note that the indices of only one of the pair of lines that makes up a band are shown. The <110> zones are 35.26° away along one of the {022} Kikuchi bands in the directions indicated (see Appendix C). Compare with *Figure 4.5a*.

at the lowest index zone, such as [001], already). When tilting the specimen in this way, it is important that the specimen is initially at the eucentric height (see Section 1.9); if you find that you have to use the minor tilt axis and you do not have software that corrects for its non-eucentric behaviour, you may have to return periodically to the image in order to relocate the area of interest, to adjust the specimen height and to focus the image. Alternatively, if spots are visible in the diffraction pattern, the diffraction pattern can be defocused so that an out-of-focus image is visible in the undeviated beam and any movement of the specimen can be compensated for with the specimen traverses.

Orienting the crystal so that the beam is along a particular zone axis has become easier now that software is available that will index diffraction patterns on-line and tilt the specimen automatically to the required axis once two other zones have been identified (see Appendix F). You should note, however, that such labour-saving devices are no substitute for understanding what is going on in reciprocal space!

4.4 Finding the exact orientation of the crystal

The Kikuchi pattern can be used to find the exact orientation of a crystal when the incident beam is not exactly along a zone axis. An example is shown in *Figure 4.6*. The diffraction pattern in *Figure 4.6a* is from a sample of aluminium in which the incident beam is close to [011]. As the zone axis is at the point at which the Kikuchi bands intersect, this point can be identified and its distance from the incident beam can be measured, as shown schematically in *Figure 4.6b*. In order to calibrate distance against angle, the spacing of a pair of Kikuchi lines (or the spacing of the corresponding spots), R, is measured; then, as $\theta = \sin^{-1}(\lambda/2d)$, θ can be calculated from the known d-spacing and electron wavelength. R is equivalent to twice the Bragg angle, 2θ and therefore the distance X in *Figure 4.6b* is equivalent to an angle of $x = X \times (2\theta_{200}/R_{200})°$.

For the example in *Figure 4.6*, $a = 0.404$ nm and therefore $d_{200} = 0.202$ nm and $\theta = 0.525°$ (for 100 kV electrons $\lambda = 0.0037$ nm, Appendix D) (*Figure 6.4a*). $R_{200} = 7.0$ mm and $X = 50$ mm; therefore, $x = 7.5°$. Similarly, the distance $Y = 6.50$ mm and the corresponding angle $y = 0.98°$. Thus, the crystal is 7.5° from [011] rotated about the normal to the (200) planes (which is the same as the [100] direction in the cubic system) plus 0.98° from [011] rotated about the normal to the (02$\bar{2}$) planes (or the [01$\bar{1}$] direction). The resultant angle of tilt, z, can be worked out from the formula (Gard, 1971):

$$\text{Tan } z = (\tan^2 x + \tan^2 y + \tan^2 x \tan^2 y)^{1/2}.$$

In this case, the resultant tilt is 7.6°. Note that there is an a 180° ambiguity in the sense of the direction of the resultant tilt, as outlined in Section 2.4.1.

For crystals that are not close to a recognizable zone axis, the procedure for finding the exact orientation is described by Williams and Carter (1996).

Exercises

4.1 Calculate the value of the deviation parameter s for the 222 reflection from the stainless steel specimen in *Figure 4.1b, c* and *d*. The lattice parameter $a = 0.3157$ nm and the wavelength was 0.0037 nm (100 kV electrons).

4.2 *Figure 4.8* is a Kikuchi pattern from a spinel which is face-centred cubic. Identify the beam direction from the symmetry of the pattern and identify the indices of the prominent Kikuchi bands. Determine the cell parameter. Note that the space group of spinel (Fd$\bar{3}$m) has, in addition to the systematic absence for the F lattice, the additional absence that for $hk0$ reflections $h + k = 4n$, where n is an integer. The

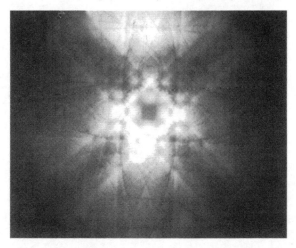

Figure 4.8. A Kikuchi pattern from a spinel which is face-centred cubic.

camera constant was 2.86 nm mm. Which Kikuchi band would you 'travel' along to reach the closest <110> zone? By what angle would you have to tilt the specimen?

4.3 *Figure 4.9* is a diffraction pattern from aluminium which is face-centred cubic, $a = 0.404$ nm. The accelerating voltage was 100 kV ($\lambda = 0.0037$ nm) The beam is close to the [011] zone direction and the ± 200 and $\pm 02\overline{2}$ Kikuchi bands are indicated. Calculate the tilt needed to bring the zone axis exactly parallel to the electron beam.

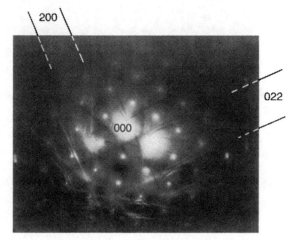

Figure 4.9. Diffraction pattern from aluminium which is face-centred cubic, $a = 0.404$ nm. The beam is close to the [011] zone direction and the ± 200 and $\pm 02\overline{2}$ Kikuchi bands are indicated.

5 The intensities of reflections

We saw in Section 3.1 that the intensity of a given diffraction spot depends upon the thickness of the crystal and the deviation of the reflection from the exact Bragg position, i.e. the exact orientation of the crystal with respect to the incident beam. Clearly, the intensity will also depend upon the types of atoms that make up the crystal and their positions in the unit cell. In this chapter, we will investigate this dependence and look into the kinematic and dynamical theories of electron diffraction.

5.1 The atomic scattering amplitude, f

In Chapters 2 and 3, we looked at diffraction in terms of reflections from lattice planes. In reality, of course, it is the atoms within the structure that diffract, or 'scatter', the electrons, and specifically the electrons and nucleus within them. The **atomic scattering amplitude**, f, is a measure of the efficiency of an isolated atom in diffracting electrons and is defined as the ratio of the scattering amplitude of the atom divided by that of a single electron; it has the dimensions of length.

The value of f falls off rapidly with scattering angle as shown in *Figure 5.1* and it does so more rapidly the larger the radius of the atom. Au and Al have about the same atomic radius (0.144 and 0.142 nm, respectively), but that of Cu is slightly smaller (0.128 nm); as a consequence, the scattering factor for Cu falls off slightly less steeply with scattering angle than those for Au and Al. Scattering is also more efficient the higher the atomic number of the atom and the longer the wavelength of the electrons (i.e. the lower the accelerating voltage).

5.2 The structure factor, F_{hkl}

The **structure factor**, F_{hkl}, is a measure of the amplitude scattered by all the atoms in a unit cell into the reflection hkl, and, like f, has

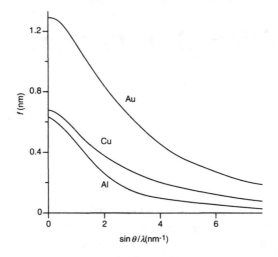

Figure 5.1. The variation of the atomic scattering factor, *f*, with (sin θ)/λ for Au (*Z* = 79), Cu (*Z* = 29) and Al (*Z* = 13), showing that elastic scattering decreases with angle away from the undeviated beam (θ = 0°) and increases with the atomic number, *Z*. *f* has dimensions of length.

dimensions of length. However, F_{hkl} is not a simple sum of all the values of *f* because we have to take into account the phase difference between the waves scattered by the individual atoms in the summation. In Section 2.2, we met the concept of the path difference, Δ, between scattered rays. The phase difference, ϕ, is a way of expressing the extent that the rays are out-of-step or out-of-phase with each other as an angle, rather than as a distance. A path difference λ between two rays is taken to be equivalent to a phase difference of 360° or 2π radians. Therefore, by proportions, a path difference of Δ is equivalent to a phase difference of $\phi = \Delta \times 2\pi/\lambda$ radians.

F_{hkl} is normally expressed as a vector or as a complex number (for an introduction to vectors and complex numbers and their use in crystallography, see Hammond, 2001):

$$F_{hkl} = \sum_{n=1}^{N} f_n(\cos \phi_n + \sin i\phi_n) = \sum_{n=1}^{N} f_n \exp i\phi_n = |F_{hkl}| \exp i\Phi_{hkl}$$

$$= |F_{hkl}|(\cos \Phi_{hkl} + i \sin \Phi_{hkl})$$

where N is the total number of atoms in the unit cell, the subscript n denotes the nth atom, $|F_{hkl}|$ is the modulus of F_{hkl}, Φ_{hkl} is the phase of the resultant reflection hkl and $i = \sqrt{-1}$. The intensity diffracted by the unit cell into the hkl reflection, I_{hkl}, is the square of the amplitude F_{hkl}:

$$I_{hkl} = F_{hkl} \times F^*_{hkl} = \left(\sum_{n=1}^{N} f_n \cos \phi_n \right)^2 + \left(\sum_{n=1}^{N} f_n \sin \phi_n \right)^2$$

where F^*_{hkl} is known as the complex conjugate of F_{hkl} and is equal to $|F_{hkl}|$ (cos Φ_{hkl} − i sin Φ_{hkl}). (If a complex number is expressed as $z = x + iy$, then its complex conjugate is $z^* = x - iy$ and $z.z^* = x^2 + y^2$.) The intensity is a scalar, rather than a vector or complex, quantity.

To evaluate F_{hkl} we need to know how to calculate ϕ_n, and hence Φ_{hkl}. I will take a semi-quantitative approach; the full mathematical treatment can be found in Hammond (2001) or Williams and Carter (1996).

Figure 5.2 shows a projection of an orthorhombic unit cell onto the (001) plane and four atoms, with the trace of the set of planes (h00) (where $h = 6$) indicated. If we assume that the atom A, for which the x coordinate is zero, scatters the electrons into the $h00$ reflection with a path difference $\Delta = 0$ and a phase angle $\phi = 0$; the atom A' in the equivalent position to A in the next unit cell, i.e. with $x = a$, will scatter the electrons with a path difference $h\lambda$ and a phase angle of $2\pi h$ radians and atom B, with x coordinate a/h on the first (h00) plane from the origin, will scatter with a path difference of λ and a phase angle 2π. By simple proportions, atom C with an arbitrary x coordinate ax_n will scatter into the $h00$ reflection with a path difference $\lambda h x_n$ and a phase angle of $\phi_n = (2\pi/\lambda).\lambda h x_n = 2\pi h x_n$. Thus, we can associate a distance of ax_n along the x-axis with a phase angle of $2\pi h x_n$. Similarly, an atom with a fractional coordinate y_n along the y-axis would scatter electrons into the $0k0$ reflection with a phase angle $2\pi k y_n$ and an atom with a fractional coordinate z_n along the z-axis would scatter electrons into the $00l$ reflection with a phase angle $2\pi l z_n$.

In the general case, the phase of electrons scattered into the hkl reflection by an atom with fractional coordinates $x_n y_n z_n$ will be:

$$\phi_n = 2\pi(hx_n + ky_n + lz_n).$$

We normally use the equation for the structure factor for the unit cell in its exponential form because this makes evaluation of the equation easier. The equation therefore becomes:

$$F_{hkl} = \sum_{n=1}^{N} f_n \exp 2\pi i(hx_n + ky_n + lz_n).$$

5.2.1 Systematic absences due to lattice type

Using the equation for F_{hkl}, we are now in a position to derive mathematically the systematic absences for a lattice type. We will take

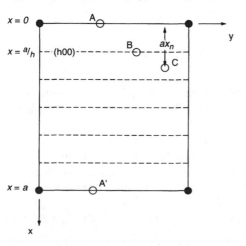

Figure 5.2. Projection of a unit cell on (001) and four atoms A, B, C and A' (the atoms A and A' are equivalent). The planes (h00) ($h = 6$) are shown in projection as dashed lines. The black dots are lattice points.

the body-centred lattice as an example (see *Figure 2.13*). For every atom at x_n, y_n, z_n there is an identical atom at $x_n + 1/2, y_n + 1/2, z_n + 1/2$ (see Appendix A). The structure factor becomes:

$$F_{hkl} = \sum_{n=1}^{N/2} f_n \left[\exp 2\pi i (hx_n + ky_n + lz_n) + \exp 2\pi i \{h(x_n + 1/2) + k(y_n + 1/2) + l(z_n + 1/2)\}\right]$$

$$= \sum_{n=1}^{N/2} f_n \left[\exp 2\pi i (hx_n + ky_n + lz_n) \times \{1 + \exp 2\pi i (h + k + l)/2\}\right].$$

Now, as $\exp n\pi i = (-1)^n$:

$$F_{hkl} = \sum_{n=1}^{N/2} f_n \left[\exp 2\pi i (hx_n + ky_n + lz_n)\right] \text{ if } h + k + l \text{ is even and}$$

$$F_{hkl} = 0 \qquad\qquad \text{ if } h + k + l \text{ is odd.}$$

So, reflections that have $h + k + l$ odd are absent, as we noted in *Table 2.1*.

5.2.2 *Crystal with a simple structure*

Caesium chloride, CsCl, is cubic and has one Cs and one Cl atom in the unit cell. If Cs is taken to be at the origin of the unit cell then Cl is at the body centre. (Note that the lattice is not body centred, however, as the atoms are not identical; it is primitive, each lattice point representing one Cs and one Cl.) If we take Cs at the origin, 000, then Cl is at 1/2, 1/2, 1/2. The structure factor is:

$$F_{hkl} = f_{Cs} \exp 2\pi i(0) + f_{Cl} \exp 2\pi i[h/2 + k/2 + l/2]$$

$$= f_{Cs} + f_{Cl} \qquad \text{if } h + k + l \text{ is even}$$

$$= f_{Cs} - f_{Cl} \qquad \text{if } h + k + l \text{ is odd.}$$

Thus, the reflections that are absent if the lattice type is truly body centred are equal to the difference in the atomic scattering factors, and therefore weak, for the primitive CsCl structure.

5.2.3 *Systematic absences due to translational symmetry elements*

If you are not familiar with translational symmetry elements and space groups, it is suggested that you read Appendix B before tackling this section.

Space-group symmetry elements that involve translation result in systematic absences, although in most cases the systematically absent spots do occur in the diffraction pattern by double diffraction (see Section 7.1.1). A c-glide plane parallel to (010) is taken as an example. For every atom at x_n, y_n, z_n there is an identical atom at $x_n, -y_n, z_n + 1/2$ (*Figure 5.3*). Substituting these two atomic positions, the structure factor becomes:

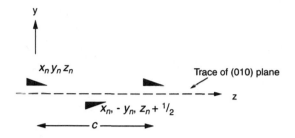

Figure 5.3. Schematic diagram showing the operation of a *c*-glide plane parallel to the (010) plane. For every atom at fractional coordinates x_n, y_n, z_n there is an identical atom at x_n, $-y_n$, $z_n + 1/2$. The dashed line is the symbol for a glide plane with translation within the plane of projection.

$$F_{hkl} = \sum_{n=1}^{N/2} f_n \left[\exp 2\pi i(hx_n + ky_n + lz_n) + \exp 2\pi i(hx_n - ky_n + l\{z_n + 1/2\})\right]$$

$$= \sum_{n=1}^{N/2} f_n \exp 2\pi i(hx_n + ky_n + lz_n)[1 + \exp 2\pi i(-2ky_n + l/2)].$$

For the case when k is zero:

$$F_{h0l} = \sum_{n=1}^{N/2} f_n \exp 2\pi i(hx_n + lz_n)[1 + \exp 2\pi i(l/2)]$$

$$= 0 \qquad \text{if } l \text{ is odd.}$$

In other words, $h0l$ reflections are absent if l is odd (*Figure 5.4*). Other systematic absences due to translational symmetry are shown in *Table 5.1*.

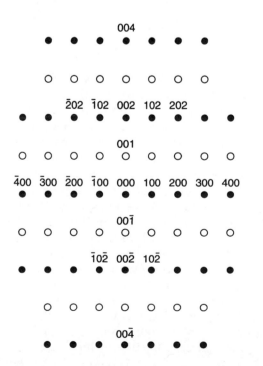

Figure 5.4. Schematic diffraction pattern of the [010] zone of an orthorhombic crystal that has a *c*-glide plane parallel to (010). All the reflections that appear in the pattern have *l* even (black dots). The systematically absent reflections are shown as open circles.

Table 5.1. Conditions for reflection for translational symmetry elements

Symmetry element	Symbol	Class of reflection	Conditions for reflection	
Glide plane // (100)	b	0kl	k	= 2n
	c		l	= 2n
	n		k + l	= 2n
	d		k + l	= 4n
Glide plane // (010)	a	h0l	h	= 2n
	c		l	= 2n
	n		h + l	= 2n
	d		h + l	= 4n
Glide plane // (001)	a	hk0	h	= 2n
	b		k	= 2n
	n		h + k	= 2n
	d		h + k	= 4n
Glide plane // (110)	c	hhl	l	= 2n
	n		l	= 2n
	d		2h + l	= 4n
Screw axis // x	$2_1, 4_2$	h00	h	= 2n
	$4_1, 4_3$		h	= 4n
Screw axis // y	$2_1, 4_2$	0k0	k	= 2n
	$4_1, 4_3$		k	= 4n
Screw axis // z	$2_1, 4_2, 6_3$	00l	l	= 2n
	$3_1, 3_2, 6_2, 6_3$		l	= 3n
	$4_1, 4_2$		l	= 4n
	$6_1, 6_5$		l	= 6n
Screw axis // [110]	2_1	hh0	h	= 2n

n = odd integer, $2n$ = even integer etc.
Note: systematic absences for lattice types are given in *Table 2.1*.

The systematic absences for all the 230 space groups are given in the *International Tables for X-ray Crystallography, Volume I* (Henry and Lonsdale, 1952) or the more recent *International Tables for Crystallography, Volume A* (Hahn, 1984). The example shown in *Figure 5.5* is the entry for $P4_2/mnm$ (full symbol $P4_2/m\ 2_1/n\ 2/m$), which is the space group of rutile (see *Figure 2.18*). The lattice is primitive and the structure has a 4_2 screw tetrad axis parallel to the z-axis with a mirror plane perpendicular to z (symbol $4_2/m$); screw diads parallel to the x- and y-axes and diagonal glide-planes perpendicular to them, (symbol $2_1/n$); diads parallel <110> and a mirror plane parallel to the {110} planes (symbol $2/m$). The right-hand column in *Figure 5.5* lists the conditions limiting possible reflections (i.e. the conditions are those that need to be fulfilled for the reflections to be present) and the first group of entries (headed General) apply to any crystal with this space group. The only condition listed is for $0kl$ reflections, for which $k + l = 2n$. This is the result of the n-glide plane parallel to the (100) plane in the space group. The x- and y-axes are symmetrically equivalent in the tetragonal system and there is therefore also an n-glide plane parallel to (010), which leads to

$P4_2/mnm$ No. 136 $P\,4_2/m\,2_1/n\,2/m$ $4/m\,m\,m$ **Tetragonal**

D_{4h}^{14}

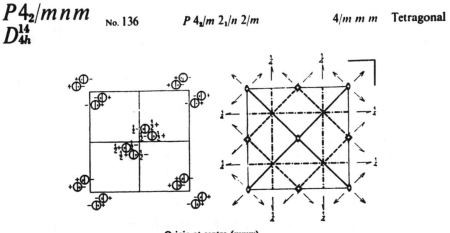

Origin at centre (*mmm*)

Number of positions, Wyckoff notation, and point symmetry			Co-ordinates of equivalent positions	Conditions limiting possible reflections
				General:
16	k	1	$x,y,z;\;\; \bar{x},\bar{y},z;\;\; \tfrac{1}{2}+x,\tfrac{1}{2}-y,\tfrac{1}{2}+z;\;\; \tfrac{1}{2}-x,\tfrac{1}{2}+y,\tfrac{1}{2}+z;$	*hkl*: No conditions
			$x,y,\bar{z};\;\; \bar{x},\bar{y},\bar{z};\;\; \tfrac{1}{2}+x,\tfrac{1}{2}-y,\tfrac{1}{2}-z;\;\; \tfrac{1}{2}-x,\tfrac{1}{2}+y,\tfrac{1}{2}-z;$	*hk0*: No conditions
			$y,x,z;\;\; \bar{y},\bar{x},z;\;\; \tfrac{1}{2}+y,\tfrac{1}{2}-x,\tfrac{1}{2}+z;\;\; \tfrac{1}{2}-y,\tfrac{1}{2}+x,\tfrac{1}{2}+z;$	*0kl*: $k+l=2n$
			$y,x,\bar{z};\;\; \bar{y},\bar{x},\bar{z};\;\; \tfrac{1}{2}+y,\tfrac{1}{2}-x,\tfrac{1}{2}-z;\;\; \tfrac{1}{2}-y,\tfrac{1}{2}+x,\tfrac{1}{2}-z.$	*hhl*: No conditions
				Special: as above, plus
8	j	m	$x,x,z;\;\; \bar{x},\bar{x},z;\;\; \tfrac{1}{2}+x,\tfrac{1}{2}-x,\tfrac{1}{2}+z;\;\; \tfrac{1}{2}-x,\tfrac{1}{2}+x,\tfrac{1}{2}+z;$	
			$x,x,\bar{z};\;\; \bar{x},\bar{x},\bar{z};\;\; \tfrac{1}{2}+x,\tfrac{1}{2}-x,\tfrac{1}{2}-z;\;\; \tfrac{1}{2}-x,\tfrac{1}{2}+x,\tfrac{1}{2}-z.$	
8	i	m	$x,y,0;\;\; \bar{x},\bar{y},0;\;\; \tfrac{1}{2}+x,\tfrac{1}{2}-y,\tfrac{1}{2};\;\; \tfrac{1}{2}-x,\tfrac{1}{2}+y,\tfrac{1}{2};$	no extra conditions
			$y,x,0;\;\; \bar{y},\bar{x},0;\;\; \tfrac{1}{2}+y,\tfrac{1}{2}-x,\tfrac{1}{2};\;\; \tfrac{1}{2}-y,\tfrac{1}{2}+x,\tfrac{1}{2}.$	
8	h	2	$0,\tfrac{1}{2},z;\;\; 0,\tfrac{1}{2},\bar{z};\;\; 0,\tfrac{1}{2},\tfrac{1}{2}+z;\;\; 0,\tfrac{1}{2},\tfrac{1}{2}-z;$	*hkl*: $h+k=2n;\; l=2n$
			$\tfrac{1}{2},0,z;\;\; \tfrac{1}{2},0,\bar{z};\;\; \tfrac{1}{2},0,\tfrac{1}{2}+z;\;\; \tfrac{1}{2},0,\tfrac{1}{2}-z.$	
4	g	mm	$x,\bar{x},0;\;\; \bar{x},x,0;\;\; \tfrac{1}{2}+x,\tfrac{1}{2}+x,\tfrac{1}{2};\;\; \tfrac{1}{2}-x,\tfrac{1}{2}-x,\tfrac{1}{2}.$	no extra conditions
4	f	mm	$x,x,0;\;\; \bar{x},\bar{x},0;\;\; \tfrac{1}{2}+x,\tfrac{1}{2}-x,\tfrac{1}{2};\;\; \tfrac{1}{2}-x,\tfrac{1}{2}+x,\tfrac{1}{2}.$	
4	e	mm	$0,0,z;\;\; 0,0,\bar{z};\;\; \tfrac{1}{2},\tfrac{1}{2},\tfrac{1}{2}+z;\;\; \tfrac{1}{2},\tfrac{1}{2},\tfrac{1}{2}-z.$	*hkl*: $h+k+l=2n$
4	d	$\bar{4}$	$0,\tfrac{1}{2},\tfrac{1}{4};\;\; \tfrac{1}{2},0,\tfrac{1}{4};\;\; 0,\tfrac{1}{2},\tfrac{3}{4};\;\; \tfrac{1}{2},0,\tfrac{3}{4}.$	*hkl*: $h+k=2n;\; l=2n$
4	c	$2/m$	$0,\tfrac{1}{2},0;\;\; \tfrac{1}{2},0,0;\;\; 0,\tfrac{1}{2},\tfrac{1}{2};\;\; \tfrac{1}{2},0,\tfrac{1}{2}.$	
2	b	mmm	$0,0,\tfrac{1}{2};\;\; \tfrac{1}{2},\tfrac{1}{2},0.$	*hkl*: $h+k+l=2n$
2	a	mmm	$0,0,0;\;\; \tfrac{1}{2},\tfrac{1}{2},\tfrac{1}{2}.$	

Figure 5.5. Symmetry information as given in the old edition (Henry and Lonsdale, 1952) of the *International Tables for Crystallography* for space group number 136, P4$_2$/*mnm* (full symbol P4$_2$/m 2$_1$/n 2/m).

the (unstated) condition that for *h0l* reflections $h + l = 2n$. The conditions for the two screw axes are not given specifically as they are conditions that are referred to as 'tied' to those for the *n*-glide planes; the screw tetrad parallel to the z-axis gives rise to the condition that for 00*l* reflections, $l = 2n$ and the screw diads parallel to the x- and y-axes give rise to the condition that for $h00$ or $0k0$ reflections, h or $k = 2n$, respectively (*Table 5.1*). All of these conditions are special cases of those stated above for the n-glide planes.

The special conditions listed below the general conditions only apply if all the atoms are in the special equivalent positions in the unit cell whose coordinates are listed in the central column.

Many of the computer programs listed in Appendix F allow you to type in the space group, and sometimes the atomic positions as well; diffraction patterns can then be generated that take account of all systematic absences and, in some cases, of double diffraction when it arises.

5.3 The kinematic and dynamical theories of electron diffraction

As we saw in Section 3.1, in addition to the contribution of the structure factor, the intensity of a diffracted beam also depends upon the deviation of the reciprocal lattice point from the exact Bragg condition, as shown in *Figure 3.6*. The intensity also depends upon the thickness of the specimen, t. If the intensity of the incident beam is equal to 1, the total variation in intensity of the diffracted beam can be expressed by the equation:

$$\Psi^2_{hkl} = \frac{\sin^2(\pi ts)}{(\xi s)^2}$$

where ξ is a material parameter called the **extinction distance** that is related to the structure factor, F_{hkl}, by the expression:

$$\xi = \pi V_c(\cos \theta)/\lambda F_{hkl}$$

where V_c is the volume of the unit cell, λ is the wavelength of the electrons and θ is the Bragg angle. ξ is a few tens of nanometers for most reflections of most inorganic materials.

The above expression for the intensity is only strictly accurate for very thin specimens because it assumes that the intensity of the diffracted beam is small in comparison with that of the incident beam. These are so-called **kinematic conditions**. In particular, the equation is clearly invalid for very small values of s because the calculated diffraction intensity then becomes equal to $(\pi t/\xi)^2$, which is greater than 1 if the thickness of the crystal is greater than ξ/π, i.e. the diffracted intensity would be greater than the intensity of the incident beam!

For crystals thicker than about one third of the extinction distance, when the intensity of the diffracted beam becomes significant and the diffracted wave can itself be re-diffracted (so-called **multiple scattering**), the **dynamical theory** is needed. The simplest form of this theory considers the interaction between the undeviated beam and one diffracted beam. In other words, this **two-beam theory** assumes that all other diffracted beams are very weak. The more comprehensive **many-beam theories** that have been developed show that the two-beam theory gives a good approximation to reality. The two-beam theory predicts that, if we define a quantity called effective deviation parameter, s' where:

$$s' = \left[s^2 + \left(\frac{1}{\xi} \right)^2 \right]^{1/2}$$

then the resultant diffracted intensity for a particular reflection from a perfect crystal of thickness t is given by:

$$\Psi^2{}_{hkl} = \frac{\sin^2(\pi t s')}{(\xi s')^2}$$

which is the same as the kinematic expression, except that s' has replaced s. Now, when s is zero, s' is equal to ξ^{-1} and $\Psi^2{}_{hkl} = \sin^2(\pi t/\xi)$. Note that these equations predict that $\Psi^2{}_{hkl}$ is periodic in t and s'. As a consequence, the intensity of the undeviated beam, $\Psi^2{}_0$ is also periodic in a complementary manner as, in order to conserve energy,

$$\Psi_0^2 = 1 - \Psi_{hkl}^2.$$

In addition to predicting the intensity of the diffracted and undeviated beams, these expressions describe features in the image, such as extinction contours and thickness fringes, very well, particularly when absorption is take into account. Further details of this subject are beyond the scope of this book, but you can find out more from Williams and Carter (1996), Thomas and Goringe (1979) or Hirsch *et al.* (1965).

5.3.1 Measuring the structure factor, F_{hkl}, and the extinction distance, ξ

Clearly, if the structure of a crystal is already known, it is a simple matter to calculate the values of F_{hkl} and ξ (tables of the values of f are given in the *International Tables for X-ray Crystallography, Volume III*, MacGillavry and Rieck, 1983, and in Hirsch *et al.*, 1965). However, if the structure is not known, it is a far from straightforward process to determine these values from conventional electron diffraction patterns of inorganic materials because of the way in which the intensities of the reflections depend upon the orientation and the thickness of the sample. In favourable circumstances, however, it is possible to measure F_{hkl} and ξ of inorganic materials by convergent-beam and other methods (see, for example, Spence and Zuo, 1992 or Spence, 1992).

Biological samples, on the other hand, are often thin enough for kinematic conditions to apply and structure factors can be measured from conventional diffraction patterns (see Misell and Brown, 1987). In favourable circumstances it has even been possible to 'construct' the crystal structures of biological materials by Fourier synthesis. This process involves mathematically 'adding together' the diffracted beams in the same way that a lens does (Section 1.1). The amplitudes of the diffracted beams, $|F_{hkl}|$ are obtained from the diffraction patterns and the phases are obtained from images of the specimen (e.g. Unwin and Henderson, 1975).

Exercises

5.1 (a) Write down the co-ordinates of an atom related to an atom in a general position x_n, y_n, z_n by a screw diad parallel to the y-axis. (b) From the expression for F_{hkl} deduce the systematic absences. (c) What is the physical reason for these absences? (d) Look at *Table 5.1* and determine what other translational symmetry elements would also lead to these absences.

5.2 The [010] zone-axis pattern in *Figure 5.4* was described in the text as an example of the effect of a c-glide plane parallel to (010). What else could produce the systematic absences seen in this section and how could the two possibilities be distinguished?

6 Determination of the Bravais lattice, point group and space group

In this chapter, we will see how electron diffraction can help us to determine the crystal system of a material, its lattice type, its point group and its space group. Most of these methods use a convergent electron beam, a technique that was introduced briefly in Chapter 1. You will need to have a good understanding of crystallography and space groups to follow the material in this chapter. If your understanding is patchy, it is recommended that you first read Appendices A and B, and possibly some of the references at the end of them.

6.1 Determination of the crystal system from conventional electron diffraction patterns

With the large tilt angles that are available in modern TEMs it is possible to examine large volumes of reciprocal space. It is, in principle, a straightforward matter to decide whether or not the crystal is cubic, hexagonal or has lower symmetry from observation of the geometries of a series of spot patterns tilted through large angles. However, it is not so easy to distinguish between the systems of lower symmetry by this method. It is important to tilt the sample to prominent zone axes (i.e. ones with low indices), as described in Section 4.3, as these will show the highest symmetry characteristic of the sample. If, for example, no diffraction patterns with four- or six-fold symmetry are found, the crystal cannot be cubic or hexagonal. The angles between prominent zones can also help to identify a crystal as cubic because, in this system, the angles are not dependent on the chemistry of the material (the angles between zones in the cubic system are given in Appendix C.3).

6.2 Convergent-beam methods

6.2.1 *Zonal repeats parallel to the electron beam*

When a convergent electron beam is used to form a diffraction pattern, the range of incident angles leads to a significant excitation of the reflections from the higher-order Laue zones or **HOLZs**. If the electron beam is exactly along a zone axis of a crystal, as shown schematically in *Figure 6.1*, it is clear that the spacing, H^*, of the reciprocal lattice layers perpendicular to the electron beam, and hence the zone-axis repeat in the crystal, can be derived from the radius, G, of the HOLZs. For the first-order Laue zone or **FOLZ**, assuming the angle α to be small, from triangle QFP: $2\alpha = \lambda G_1$ radians and from triangle FOP: $\alpha = H^*/G_1$. Equating the two expressions for α gives:

$$H^* = G_1^2 \lambda/2.$$

Similarly, for the second-order Laue zone, or SOLZ:

$$H^* = G_2^2 \lambda/4.$$

The spacing of the reciprocal layers is given by:

$$H^* = N/r_{[UVW]}$$

where $r_{[UVW]}$ is the zonal repeat and the value of N takes account of systematic absences due to lattice type (Section 2.7). N is either 1, 2 or 3, as set out below.

- For P or R lattices: $N = 1$ for any axis.
- For F lattices: $N = 2$ if $U + V + W$ is even, otherwise $N = 1$.
- For I lattices: $N = 2$ if U, V, W are all odd, otherwise $N = 1$.
- For C lattices: $N = 2$ if U and V are both odd, otherwise $N = 1$.
- For R lattices where hexagonal indices have been used: $N = 3$ if $U - V + W \neq 3n$, where n is an integer, otherwise $N = 1$.

From the FOLZ ring the zonal repeat is therefore given by:

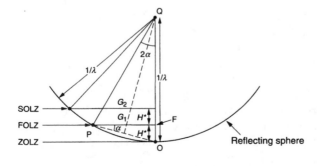

Figure 6.1. Diagram illustrating the intersection of Laue zones with the reflecting sphere when the electron beam is parallel to a zone axis. The zero-, first- and second-order Laue zones (called the ZOLZ, FOLZ and SOLZ, respectively) are shown.

$$r_{[UVW]} = 2N/(\lambda G_1^2) = \left(\frac{2N}{\lambda}\right)\left(\frac{\lambda L}{R_1}\right)^2$$

where R_1 ($= \lambda L G_1$) is the radius of the FOLZ measured on the film and λL is the camera constant. Similarly, from the SOLZ ring:

$$r_{[UVW]} = 4N/(\lambda G_2^2) = \left(\frac{4N}{\lambda}\right)\left(\frac{\lambda L}{R_2}\right)^2.$$

In terms of the unit-cell parameters, the value of the zonal repeat is:

$$r^2_{[UVW]} = U^2 a^2 + V^2 b^2 + W^2 c^2 + 2VWbc \cos \alpha + 2WUca \cos \beta + 2UVab \cos \gamma$$

HOLZs are best seen in patterns recorded at a short camera length and R is most easily measured if a convergent beam with a moderate to large convergence angle is used, i.e. conditions under which the diffracted beams overlap (Kossel pattern). This is because the intensity in an individual HOLZ then appears as one or more continuous rings; compare *Figure 6.2a* and *b*. When measuring R, you should measure the innermost ring of the set if there is more than one. It is also better to measure the FOLZ rather than a higher-order ring because, if the scattering angle is too large ($\alpha \sim 10°$), the measurements may suffer as a result of lens distortions. The effect of the lens distortion in reciprocal space should be measured using a standard specimen with an accurately known zonal repeat parallel to the electron beam. The distance R and the camera constant λL should be measured as accurately as possible in order to minimize the errors in H^*, as both quantities are squared in the equation

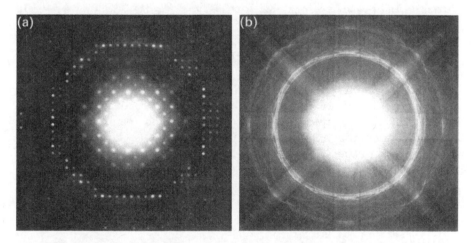

Figure 6.2. [001] CBED patterns from the spinel gahnite, $ZnAl_2O_4$, which is cubic with space group $Fd\bar{3}m$ (full symbol $F4_1/d\,\bar{3}\,2/m$). Patterns were taken at an original camera length of 210 mm in a Philips 400T electron microscope. (a) Pattern taken under microdiffraction conditions in the microprobe mode with a 30 μm condenser aperture. Individual reflections can be seen in the ZOLZ and the FOLZ. (b) The same pattern recorded under Kossel conditions in the nanoprobe mode with a 100 μm condenser aperture. The FOLZ now consists of a series of closely spaced, continuous rings.

for $r_{[UVW]}$. (Note that an error in λL gives a % error that is different in the HOLZ from in the zero-order Laue zone, or ZOLZ.) The diffraction astigmatism should also be corrected carefully (Section 1.11).

The measurement of the zonal repeat parallel to the electron beam allows the presence of **polytypes** (materials with the same chemistry, but different stacking sequences, and hence different lattice repeats, along one axis) to be detected. It may also allow the distinction between cubic and tetragonal crystals to be made if the diffraction pattern is viewed along [001]. In this latter case, the crystal system can be determined from a set of diffraction patterns from a single axis.

6.2.2 Geometry of CBED patterns from HOLZs

The crystal system and lattice type can be determined from the **positions** of the reflections in the HOLZ compared with those in the ZOLZ. Zone-axis patterns, or ZAPs, need to be recorded in specific directions of high symmetry (Morniroli and Steeds, 1992). In addition, differences between the **periodicity** of reflections in the ZOLZ and the FOLZ indicate the presence of a glide plane perpendicular to the electron beam. The visibility of reflections in HOLZs may be enhanced by the use of a stage cooled by liquid nitrogen or by reducing the operating voltage (atoms scatter lower-energy, longer-wavelength electrons more strongly, see *Figure 5.1*). If the FOLZ reflections still cannot be seen, the crystal should be tilted along each of the symmetry elements in the ZOLZ net until reflections belonging to the FOLZ appear, as shown in *Figure 3.5*.

Finding the crystal system and lattice type. *Figure 6.3a* shows a [010] microdiffraction pattern of synthetic $MgGeO_3$ pyroxene (a single-chain silicate) in which individual spots in the ZOLZ and FOLZ can be seen; in *Figure 6.3b*, the positions of the reflections relative to each other are shown diagrammatically. The ZOLZ is an oblique, two-dimensional net and, from the extension of the x*- and z*-axes into the FOLZ, you can see that the FOLZ stacks vertically above the ZOLZ, except that it is displaced by half a repeat in the x* direction. The reciprocal lattice is therefore monoclinic in shape and it is centred on the C (x*–y*) face; for reflections $h0l$ in the ZOLZ, $h = 2n$, where n is an integer, whereas in the FOLZ, for reflections $h1l$, $h = 2n + 1$ (see *Figure G.2b* for a diagram of a C-face-centred reciprocal lattice). These observations show that the crystal is C-face centred monoclinic, the y-axis or [010] direction being the unique diad axis in the crystal class.

Detecting glide planes by comparing the ZOLZ and the FOLZ. In *Figure 6.3*, the spacing of the reflections along the z*-axis is halved in the FOLZ compared with the spacing in the ZOLZ. This is the result of the presence of a c-glide plane parallel to (010) in the space group as this type of glide plane results in $h0l$ reflections being absent if l is odd see (*Table 5.1*). Thus, for the ZOLZ, where all the reflections have indices $h0l$, $l = 2n$,

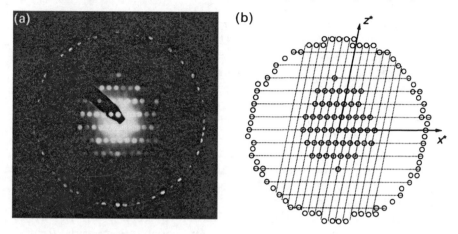

Figure 6.3. (a) [010] microdiffraction pattern of synthetic $MgGeO_3$ pyroxene taken at an original camera length of 290 mm with a divergence angle of 1.4×10^{-3} radians, 120 kV. Individual reflections can be seen in the ZOLZ and FOLZ. (b) Diagram of the pattern in (a). The superposition of the ZOLZ net (shown by the dotted lines) and its extension into the FOLZ shows that the FOLZ stacks vertically above the ZOLZ, except that it is displaced by half the a^* repeat. The spacing of the reflections along the z^*-axis is also halved in the FOLZ compared with that in the ZOLZ. From this one pattern, it can be concluded that the crystal is monoclinic and that the partial diffraction symbol is $C1c1$. The space group is either $C2/c$ or Cc. Reproduced from Champness PE, *Mineral. Mag.*, 1987, with the permission of the Mineralogical Society, London.

whereas for the FOLZ, where the indices are $h1l$, there are no restrictions on the l index.

Another example of a glide plane that can be detected by comparison of the geometry of the ZOLZ and the FOLZ is shown in the [001] pattern from spinel in *Figure 6.2a*. The reflections in the FOLZ have a spacing along x* and y* that is half the value of the spacing in the ZOLZ. This difference is the result of the d-glide plane perpendicular to the z-axis (parallel to the plane of the projection) in the space group $Fd\bar{3}m$ (full symbol $F4_1/d\ \bar{3}\ 2/m$) to which spinel belongs. The systematic absences resulting from this symmetry element occur for $hk0$ reflections when $h + k \neq 4n$, where n is an integer (see *Table 5.1*). Also, because of the F lattice, h and k must be even in the ZOLZ, where $l = 0$ (h, k and l must be all odd or all even, see *Table 2.1*). In the FOLZ, the reflections have indices $hk1$ and the only restrictions on the indices are that h and k must be odd (again because of the F lattice).

Finding the partial extinction symbol. The information that can be obtained from microdiffraction patterns about lattice type and glide planes can be combined in a **partial extinction symbol** (Morniroli and Steeds, 1992), which looks very like a point-group or space-group symbol (Appendices A and B) and can be used as a first step in finding the full point group or space group. The complete partial extinction symbol for monoclinic crystals can be obtained from the [010] microdiffraction pattern alone. The symbol for the pyroxene in *Figure 6.3* is *C*1*c*1. The symbol *C* indicates the lattice type and the next three symbols refer to the x-, y- and z-axes, respectively. The symbol 1 in the second and fourth positions indicates that there is no symmetry parallel to the x- and z-axes (because the crystal is known to be monoclinic) and the *c* in the third position indicates the *c*-glide plane perpendicular to the y-axis. The two possible space groups that would give rise to this partial extinction symbol are *C*2/*c* and *Cc*. The reason why these two space groups cannot be distinguished by means of microdiffraction patterns is that the diad axis in space group *C*2/*c* does not lead to any systematic absences in the reflections. However, we will see in Section 6.2.3 that the presence of a diad, or the lack of it, can be detected by other CBED methods.

Although a single [010] microdiffraction pattern is sufficient to determine the partial extinction symbol for monoclinic crystals, for the cubic system both <001> and <011> patterns are needed. The partial extinction symbol for the <001> pattern from the spinel in *Figure 6.2* is either *Ia*.. or *Fd*.. (*Figure 6.4a*). Once again, the first symbol is the lattice type and the second refers to the glide plane perpendicular to the x-, y- and z-axes (which are equivalent in the cubic system). The two dots that take up the third and fourth positions indicate that nothing can be concluded from the <001> pattern about the symmetry along the <111> and <011> directions, respectively. In other words, the pattern could have been produced by an I lattice with an *a*-glide plane perpendicular to the beam or by an F lattice with a *d*-glide plane perpendicular to the beam. The two possible <011> microdiffraction patterns corresponding to the <001> pattern in *Figure 6.4a* are shown diagramatically in *Figure 6.4b* and *c* and are quite distinctive. A microdiffraction pattern of the spinel taken with the electron beam along the <011> axis is shown in *Figure 6.4d*; clearly, the pattern matches *Figure 6.4c* and the partial diffraction symbol is *Fd*... The two possible space groups corresponding to this partial diffraction symbol are *Fd*$\bar{3}$*m* and *Fd*$\bar{3}$.

Morniroli and Steeds (1992) provide tables and theoretical microdiffraction patterns that allow the partial diffraction symbol to be determined from experimental microdiffraction patterns for all the crystal systems except triclinic. The patterns may also be drawn with the aid of computer software (see Appendix F).

6.2.3 *The symmetry of CBED patterns*

So far, we have only considered the information that can be obtained from the positions of the reflections in ZAPs. However, there is a wealth of detail *within* and *between* the diffracted discs of a CBED pattern recorded at the exact zone axis that can be used to provide information about the point group and space group of the crystal. Although a detailed treatment of CBED methods is beyond the scope of this book, I hope, by way of some examples, to give you an idea of what information it is possible to derive from CBED and how it is obtained. Further details about theory and

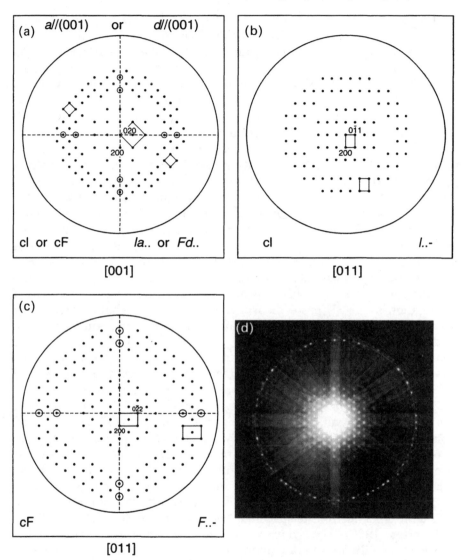

Figure 6.4. (a-c) Some schematic microdiffraction patterns for the cubic system adapted from Morniroli and Steeds (1992). The partial diffraction symbol is shown in the bottom right-hand corner in each case, the Bravais lattice is given in the bottom left-hand corner (c stands for cubic) and the glide plane perpendicular to the electron beam is given at the top. The reciprocal unit cells in the ZOLZ and the FOLZ are indicated by rectangles so that their relative sizes are clear. The circled spots in the FOLZ and the dashed lines through them indicate rows that are aligned with those in the ZOLZ. (a) <001> zone-axis pattern for the partial diffraction groups *Ia..* or *Fd...*. (b) The <011> zone-axis pattern for the partial diffraction group *I..-*. (c) The <011> zone-axis pattern for the partial diffraction group *F..-* . The dash in the fourth position of the symbols in (b) and (c) indicates that there is no glide plane perpendicular to the <011> directions. (d) <011> microdiffraction pattern from the spinel in *Figure 6.2*. It matches the schematic pattern in (c) rather than (b), so the partial diffraction symbol is *Fd...*.

methodology can be found in Eades (1988a, 1988b, 1992), Loretto (1994), Spence and Zuo (1992) and Williams and Carter (1996).

It is vitally important for the applications described below that the zone axis of interest is exactly parallel to the electron beam. It is usually most convenient do the preliminary sample tilting with a large spot size so that there is plenty of illumination on the screen and Kikuchi lines can be more easily seen. The final stage of centring is difficult to accomplish with the goniometer alone, especially if the crystal of interest is small. Rather than tilt the specimen, the electron beam can be tilted to line up with the zone axis of the crystal. This operation misaligns the microscope (making you unpopular with the next user!) and it must be re-aligned before good images can be obtained. However, the misalignment does not affect the diffraction pattern. The electron beam can be tilted in one of two ways: the condenser aperture can be moved mechanically or, alternatively, the microscope can be operated in the dark-field mode (see Section 3.3.2) and the dark-field tilt controls can be used to tilt the beam. The latter method is preferred as you can easily re-align the microscope by reverting to the bright-field mode. A further practical point: there is a wealth of information in CBED patterns over a broad range of angles (up to 5° or so) and over a broad range of intensities. All this information cannot be recorded in a single film exposure, both because of the scale of the information and because the intensity range is beyond that of any film. Therefore, in recording CBED patterns on film for the purposes of determining crystal symmetry, several exposures are needed.

The detail seen in ZAPs when the specimen is sufficiently thick is of two basic kinds: broad, diffuse features within the discs (*Figures 6.5b* and *6.6a*) and sharp lines both within the discs and outside them (*Figures 6.2b, 6.6a* and *6.6b*) The broad contrast in a ZAP arises from dynamic interactions within the ZOLZ (I will call it zero-order information). Because it is associated with quite short extinction distances (Section 5.3.1), it is sensitive to thickness and can, in favourable instances, be used to find the thickness of the specimen (see Spence and Zuo, 1992; Williams and Carter, 1996). The symmetry of the zero-order information is the projected, two-dimensional symmetry of the crystal along the zone axis, and belongs to one of the 10 two-dimensional point groups (*Table 6.1* and *Figure 6.7*).

The sharp lines within the discs arise from three-dimensional diffraction: they are the result of elastic scattering by the planes corresponding to reflections in the HOLZs. These **HOLZ lines** (sometimes called **deficit lines**) are sensitive to very small changes in lattice parameters and can, in principle, be used to measure the latter to an precision of $\sim 0.2\%$ by computer simulation (see Spence and Zuo, 1992; Williams and Carter, 1996). A corollary of this property is that small lattice strains degrade the visibility of the HOLZ lines. The HOLZ lines occur in pairs (like Kikuchi lines) with a bright (excess) line associated with the HOLZ disc responsible for it and a parallel, dark line present in the disc from the undeviated beam (the **bright-field disc**). The dark

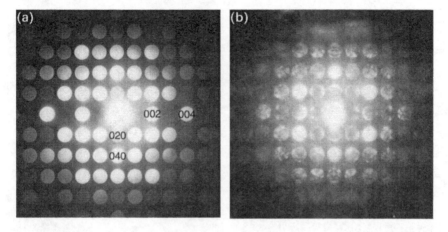

Figure 6.5. [100] microdiffraction patterns from the mineral olivine, $(Mg,Fe)_2SiO_4$. (a) The specimen is sufficiently thin that no dynamic detail is seen in the discs. (b) Microdiffraction pattern from a thicker area of the same specimen. The discs contain broad, fuzzy features that arise from dynamic interactions within the ZOLZ. In (a) the reflections 00*l* are genuinely absent when *l* is odd. This absence is the result of an *n*-glide plane parallel to (010) and a screw diad parallel to the z-axis in the space group *Pbnm* (= $P\,2_1/b\,2_1/n\,2_1/m$). In (b) the 'forbidden' 00*l* reflections are present, but contain a line of dynamic absence or GM line parallel to z* which indicates that the intensity in these reflections appears by double diffraction. Note that every other reflection along y* is systematically absent because of a *b*-glide plane parallel to (100); these reflections cannot reappear by double diffraction from within the ZOLZ. Reproduced from Champness PE, *Mineral. Mag.*, 1987, with the permission of the Mineralogical Society, London.

lines outside the discs are the result of inelastic scattering from the HOLZ planes. They are the same as the Kikuchi lines that arise from ZOLZ planes in both conventional, SAD and in CBED and are known as **HOLZ Kikuchi lines**. They are continuous with the HOLZ lines within the discs, but are generally a little less sharp.

Because extinction distances for HOLZ reflections are long, typically several micrometres, you will need a thick crystal (several extinction distances) if you want to see sharp HOLZ lines. However, the radius of the HOLZ ring is the most important factor affecting visibility (the smaller the better); the perfection of the crystal is also important. *Figures 6.2b* and *6.6b* show HOLZ Kikuchi lines, together with (ZOLZ) Kikuchi lines and excess HOLZ lines. These figures were taken from thick crystals at short camera lengths and moderate to large convergence angles, 2α, the best conditions for observing these features.

Identification of phases from ZAPs. As the HOLZ detail is associated with very long extinction distances, it is relatively insensitive to the thickness of the sample. ZAPs from high-symmetry directions can therefore be used as an aid to identification of phases. An atlas of CBED patterns for common alloy phases has been published by Steeds *et al.* (1984). CBED can distinguish between polymorphs (phases with the same chemical formula, but a different crystal structure) and compounds

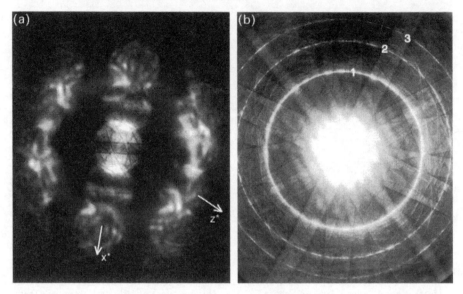

Figure 6.6. [010] CBED patterns of clinopyroxene $MgGeO_3$ taken at two different camera lengths, but the same moderate value of 2α. (a) Pattern taken at a long camera length (1150 mm) showing deficient HOLZ lines in the bright-field disc and zero-order scattering in the other ZOLZ discs. Both the BF (full) and WP (proj) symmetries are 2. (b) Pattern taken at an original camera length of 290 mm showing Kikuchi bands and excess HOLZ lines. The WP (full) symmetry is also 2. The first-, second- and third-order Laue zones are labelled 1, 2 and 3, respectively. Reproduced from Champness PE, *Mineral. Mag.*, 1987, with the permission of the Mineralogical Society, London.

that contain light elements, such as the various borides of molybdenum, which EDX cannot do (Steeds *et al.*, 1984).

Determination of the point group from ZAPs. There are four pieces of information contained in ZAPs that can help you to determine the point group of your material (Eades, 1988a), although, because of variations in experimental conditions, it will not always be possible to determine all of them. The pieces of information are as follows.

- The bright-field symmetry, i.e. the symmetry of the sharp HOLZ lines within the 000 disc; here called BF (full).
- The projection bright-field symmetry, i.e. the symmetry of the broad features within the 000 disc; here called BF (proj).
- The symmetry of the whole pattern, i.e. the symmetry of the HOLZ lines and HOLZ Kikuchi lines in the pattern as a whole; here called WP (full).
- The projection symmetry of the whole pattern, i.e. the symmetry of the broad features in the zero-layer discs; here called WP (proj).

The first step in finding the point group is to find what is known as the **diffraction group** of the material. There are 31 of these (*Tables 6.2* and *6.3*) and they describe the full three-dimensional symmetry of a CBED

Table 6.1. The relationship between the projection diffraction groups and the 2D symmetries of CBED patterns

Projection diffraction group	Symmetry of 2D information	
	WP (proj)	BF (proj)
1_R	1	2
21_R	2	2
$m1_R$	m	$2mm$
$2mm1_R$	$2mm$	$2mm$
31_R	3	6
$3m1_R$	$3m$	$6mm$
41_R	4	4
$4mm1_R$	$4mm$	$4mm$
61_R	6	6
$6mm1_R$	$6mm$	$6mm$

After Eades, 1988a.

Table 6.2. The relationship between the 31 diffraction groups and the full (3D) symmetries of CBED patterns

Diffraction group	Symmetry of high-order (3D) information		Projection diffraction group
	WP (full)	BF (full)	
1	1	1	1_R
1_R	1	2	1_R
2	2	2	21_R
2_R	1	1	21_R
21_R	2	2	21_R
m_R	1	m	$m1_R$
m	m	m	$m1_R$
$m1_R$	m	$2mm$	$m1_R$
$2m_Rm_R$	2	$2mm$	$2mm1_R$
$2mm$	$2mm$	$2mm$	$2mm1_R$
2_Rmm_R	m	m	$2mm1_R$
$2mm1_R$	$2mm$	$2mm$	$2mm1_R$
3	3	3	31_R
31_R	3	6	31_R
$3m_R$	3	$3m$	$3m1_R$
$3m$	$3m$	$3m$	$3m1_R$
$3m1_R$	$3m$	$6mm$	$3m1_R$
4	4	4	41_R
4_R	2	4	41_R
41_R	4	4	41_R
$4m_Rm_R$	4	$4mm$	$4mm1_R$
$4mm$	$4mm$	$4mm$	$4mm1_R$
4_Rmm_R	$2mm$	$4mm$	$4mm1_R$
$4mm1_R$	$4mm$	$4mm$	$4mm1_R$
6	6	6	61_R
6_R	3	3	61_R
61_R	6	6	61_R
$6m_Rm_R$	6	$6mm$	$6mm1_R$
$6mm$	$6mm$	$6mm$	$6mm1_R$
6_Rmm_R	$3m$	$3m$	$6mm1_R$
$6mm1_R$	$6mm$	$6mm$	$6mm1_R$

pattern. Possible point groups can then be found from *Figure 6.8*, as will be shown later. In favourable cases, the ZAPs from only one high-symmetry zone axis are needed to find the point group. Note that it is

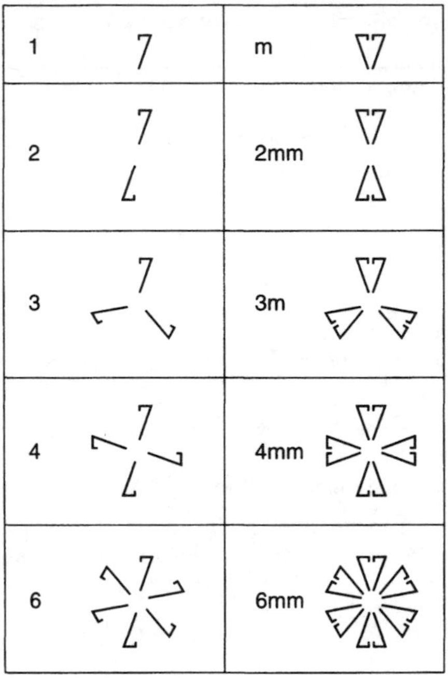

Figure 6.7. Illustration of the 10 two-dimensional crystallographic point groups. Reproduced from O'Keefe M and Hyde BG, *Crystal Structures I: Patterns and Symmetry*, 1996, with permission of the Mineralogical Society of America.

unnecessary, for the purposes of using *Tables 6.1* and *6.2* and *Figure 6.8* to find the point group, to understand the meaning of the symbols used for the diffraction groups and projection diffraction groups.

The internal symmetry of dark-field reflections at the exact Bragg condition may also be used to help determine the possible diffraction group, as described by Buxton *et al.* (1976) and Steeds (1979, 1984). However, although the additional use of the dark-field symmetry will uniquely determine the diffraction group for a ZAP (whereas the use of the four pieces of information described above may not), it is often easier and quicker to tilt the crystal to another zone axis or to use other available information about the sample (such as that described in Section 6.2.2) in order to determine its point group.

I will illustrate the use of *Tables 6.1* and *6.2* and *Figure 6.8* with an example. *Figure 6.6* shows two [010] ZAP patterns of the MgGeO$_3$ pyroxene of *Figure 6.3* taken at two different camera lengths. *Figure 6.6a* shows the BF (full) and the WP (proj) information; *Figure 6.6b* shows the WP (full) symmetry. (Note that, although the WP (full) symmetry is present in the ZOLZ, the HOLZ detail is often weak, as here, and it is usually better to record a pattern showing HOLZ reflections in order to determine the WP symmetry.) All three available pieces of information show diad (two-fold) symmetry. From *Table 6.2*, the diffraction group is either 2 or 21$_R$ and the projection diffraction group is 21$_R$. *Table 6.1* confirms these findings. *Figure 6.8* shows that these diffraction groups are consistent with point groups 2/*m*, $\overline{3}$*m* (full symbol $\overline{3}$2/*m*), 2 and 32. But, as we found from the microdiffraction patterns that the lattice is

Table 6.3. The diffraction group symmetry seen in CBED patterns from different zone axes

Point group	$<111>$	$<100>$	$<110>$	$<UV0>$	$<UUW>$	$<UVW>$
$\bar{m}3m$	$6_R mm_R$	$4mm1_R$	$2mm1_R$	$2_R mm_R$	$2_R mm_R$	2_R
$\bar{4}3m$	$3m$	$4_R mm_R$	$m1_R$	m_R	m	1
432	$3m_R$	$4m_R m_R$	$2m_R m_R$	m_R	m_R	1

Point group	$<111>$	$<100>$	$<UV0>$	$<UVW>$
$m\bar{3}$	6_R	$2mm1_R$	$2_R mm_R$	2_R
$2\bar{3}$	3	$2m_R m_R$	m_R	1

Point group	$[001]$	$<100>$	$<1\bar{1}0>$	$<UV0>$	$<UUW>$	$<U\bar{U}W>$	$<UVW>$
$6/mmm$	$6mm1_R$	$2mm1_R$	$2mm1_R$	$2_R mm_R$	$2_R mm_R$	$2_R mm_R$	2_R
$\bar{6}m2$	$3m1_R$	$m1_R$	$2mm$	m	m	m	1
$6mm$	$6mm$	$m1_R$	$m1_R$	m_R	m	m	1
622	$6m_R m_R$	$2m_R m_R$	$2m_R m_R$	m_R	m_R	m_R	1

Point group	$[001]$	$<UV0>$	$<UVW>$
$6/m$	61_R	$2_R mm_R$	2_R
$\bar{6}$	31_R	m	1
6	6	m_R	1

Point group	$[001]$	$<100>$	$<U\bar{U}W>$	$<UVW>$
$\bar{3}m$	$6_R mm_R$	21_R	$2_R mm_R$	2_R
$3m$	$3m$	1_R	m	1
32	$3m_R$	2	m_R	1

Point group	$[001]$	$<UVW>$
$\bar{3}$	6_R	2_R
3	3	1

Point group	$[001]$	$<UV0>$	$<UVW>$
$4/m$	41_R	$2_R mm_R$	2_R
$\bar{4}$	4_R	m_R	1
4	4	m_R	1

Point group	$[001]$	$<100>$	$<110>$	$<U0W>$	$<UV0>$	$<UUW>$	$<UVW>$
$4/mmm$	$4mm1_R$	$2mm1_R$	$2mm1_R$	$2_R mm_R$	$2_R mm_R$	$2_R mm_R$	2_R
$\bar{4}2m$	$4_R mm_R$	$2m_R m\ _R$	$m1_R$	m_R	m_R	m	1
$4mm$	$4mm$	$m1_R$	$m1_R$	m	m_R	m	1
422	$4m_R m_R$	$2m_R m_R$	$2m_R m_R$	m_R	m_R	m_R	1

Point group	$[001]$	$[100],[010]$	$<U0W>$	$<UV0>$	$<UVW>$
mmm	$2mm1_R$	$2mm1_R$	$2_R mm_R$	$2_R mm_R$	2_R
$mm2$	$2mm$	$m1_R$	m	m_R	1
222	$2m_R m_R$	$2m_R m_R$	m_R	m_R	1

Point group	$[010]$	$<U0W>$	$<UVW>$
$2/m$	21_R	$2_R mm_R$	2_R
m	1_R	m	1
2	2	m_R	1

Point group	$<UVW>$
$\bar{1}$	2_R
1	1

After Buxton *et al.*, 1976.

monoclinic C and the space group is either *C2/c* or *Cc*, the correct space group is *C2/c* (point group $2/m$). This example shows the power of convergent-beam methods. From one set of diffraction patterns taken along the [010] axis the space group has been determined. A conventional electron diffraction pattern of this axis will give the *a*, *c* and β cell parameters and the radius of the [010] HOLZ rings can be used to find the *b* parameter (Exercise 6.2).

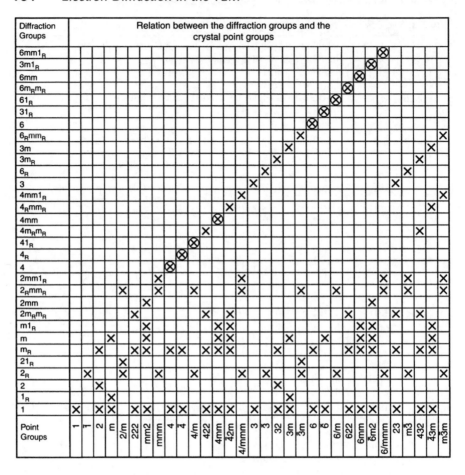

Figure 6.8. Relationship between the diffraction groups and the crystal point groups. Note that the higher the symmetry of the diffraction group (i.e. the higher it is in the table), the fewer are the possibilities for the point group. In fact, 11 diffraction groups uniquely define the point group (circled). Reproduced from Buxton *et al., Phil. Trans. Roy. Soc.*, 1976; 281A: 171–193, with permission of the Royal Society.

Determining space groups from dynamic absences. As we shall see in Section 7.1.1, certain reflections (those that should be absent in an electron diffraction pattern because the space group contains a translational symmetry element) reappear by double diffraction (Section 7.1) if the crystal is thicker than about 5 nm. The symmetry elements involved are (1) glide planes parallel to the electron beam that have a component of translation perpendicular to the beam and (2) screw diad axes that are perpendicular to the electron beam. Fortunately, when such doubly diffracted spots are recorded under convergent-beam conditions at an exact zone axis, they show a characteristic central line of missing intensity, the so-called **dynamic absence** or **GM line** (Gjønnes and Moodie, 1965; *Figures 6.5b* and *6.9*). The lines of absence are present for all sample thicknesses, occur at all operating voltages and become wider

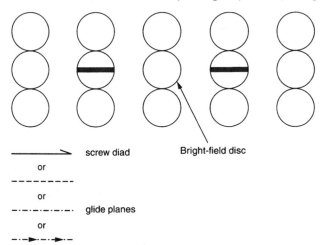

Figure 6.9. Schematic diagram of the appearance of lines of dynamic absence (represented by the heavy horizontal lines) in a CBED pattern. The GM lines occur in every other diffracted disc in a row containing the bright-field disc.

as the thickness of the specimen becomes smaller. In the kinematic limit (i.e. a very thin crystal), the line of absence fills the reflection completely; the reflection is truly absent, as in *Figure 6.5a*. The lines of absence occur along a principal axis of the ZAP and are parallel to the **g**-vectors of the reflections, which is also the direction of the translation causing the dynamic absence. When a forbidden reflection is tilted to be exactly at the Bragg angle, and if three-dimensional effects (i.e. HOLZ lines) are relatively weak, a second black line, perpendicular to the first, is seen in the disc, forming a **black cross**.

As we have seen, lines of dynamic absence or black crosses indicate either that there is a glide plane parallel to the beam, with a glide component perpendicular to it, or a screw diad axis perpendicular to the beam. Note that the 4_1, 4_3, 6_1, 6_3 and 6_5 screw axes all include the 2_1 operation (see *Figure B.1*). If the diffraction group of the zone axis is known, you can deduce the significance of the GM line, i.e. whether it is the result of a screw axis, a glide plane, or both (*Table 6.4*).

We can use *Figure 6.5b* as an example of the use of *Table 6.4*. The diffraction group for the [100] zone axis of olivine (point group *mmm*) is $2mm1_R$ (*Table 6.3*). The single row of GM lines in the $00l$ reflections for which l is odd is therefore the result both of a glide plane parallel to (010) and of a screw diad parallel to the z-axis. The glide plane is an *n*-glide (i.e. $(a + c)/2$) as the space group is *Pbnm* ($=P\ 2_1/b\ 2_1/n\ 2_1/m$). The identity of the glide plane could be determined by tilting the crystal to the [010] zone axis and recording a microdiffraction pattern. The reflections in the ZOLZ would show absences for $h + l \neq 2n$, whereas the FOLZ would show no absences.

If the point group of a material is known, the procedure described above will uniquely identify 181 of the 230 space groups. Techniques that can be used to distinguish all but four of the others are described by Eades (1988b).

Table 6.4. Deduction of the origin of GM lines in zero-layer reflections from the diffraction group

Single row of zero-layer GM lines	Perpendicular rows of zero-layer GM lines	Cause
m_R $2m_R\,m_R$	$2m_R m_R$ $4m_R m_R$	Screw axis parallel to each row of GM lines
m $2mm$	$2mm$ $4mm$	Glide plane parallel to the zone axis and each row of GM lines
$2_R mm_R$	$4_R mm_R$	Glide plane if parallel to WP (full) mirror *or* screw axis if perpendicular to WP (full) mirror
	$2_R mm_R$	Glide plane parallel *and* screw axis perpendicular to WP (full) mirror
$m1_R$ $2mm1_R$	$2mm1_R$ $4mm1_R$	Glide plane *and* screw axis parallel to each row of GM lines*

Adapted from Eades, 1988b.
*This is the case for double diffraction routes within the zero layer. If the GM lines are produced by double diffraction *via* HOLZ reflections, there are space groups for which the GM lines can be produced by a glide plane or a screw axis alone.

6.2.4 The Tanaka method

In microdiffraction patterns, such as those shown in *Figures 6.5* and *6.6a*, the maximum size of each diffraction disc, i.e. the angular field of view, is determined by the Bragg angle as the discs should not overlap significantly. For materials with large unit cells (small distances between spots in the diffraction pattern) this can be a severe limitation if the object is to determine symmetry information from the pattern. A number of methods have been devised to overcome this problem (see Spence and Zuo, 1992; Tanaka and Terauchi, 1985), but the only one that does not require a modification to the microscope is the defocus method due to Tanaka *et al.* (1980). In this method, the electron beam is focused to a probe as in a normal CBED pattern but, in this case, the focus is not in the plane of the specimen but in the plane of the diffraction aperture (the first image plane) (*Figure 6.10a*). This plane will contain a spot diffraction pattern (*Figure 6.10b*) and the diffraction aperture can be used to isolate one of the diffracted beams (*Figure 6.10c*). The microscope is then switched to the diffraction mode and a single disc of the CBED pattern will be seen on the screen of the microscope (*Figure 6.11*).

The method recommended by Eades (1984) for obtaining a Tanaka pattern for the bright-field disc is as follows.

- Start with a normal condenser aperture and choose a convenient magnification.
- Set the specimen at the eucentric height and focus the image.
- Overfocus the objective lens by the desired amount. This is likely to be tens of micrometres (see *Table 6.5*). Note: the image could be underfocused, but overfocus seems to give better results.

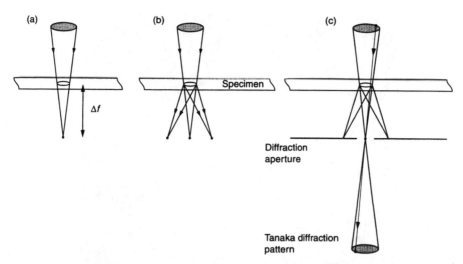

Figure 6.10. Schematic diagrams to illustrate the formation of a Tanaka CBED pattern. (a) The electron probe is focused a distance Δ*f* below the specimen, in the plane of the diffraction aperture (the first image plane of the objective lens). (b) A spot diffraction pattern is formed in the plane where the probe is focused. (c) A diffraction aperture isolates one of the diffracted beams and a single disc of the CBED pattern will be seen on the screen of the microscope when it is switched to the diffraction mode. The ray that is drawn in shows that each point in the Tanaka pattern comes from a separate part of the specimen.

- Introduce a diffraction aperture small enough to isolate the undeviated beam and centre it. This aperture is likely to be 5 or 10 μm in diameter, considerably smaller than those used for normal SAD.
- Switch to the diffraction mode and remove the condenser aperture or exchange it for a larger one. The larger the aperture the larger is the angular field of view. It is preferable to use a large aperture rather than none because the latter procedure introduces spherical aberration.

To obtain a Tanaka pattern from a dark-field disc, the same procedure is followed, but the diffraction aperture is moved so that the required diffracted beam passes along the optic axis of the microscope (Section 3.3.2). There is no need to tilt the illumination.

One disadvantage of the Tanaka method compared with the standard method of forming a CBED pattern is that, because the probe is not focused onto the specimen, the pattern comes from a larger area than normal and each part of the pattern is obtained from a different part of the specimen (as shown by the diffracted ray in *Figure 6.10c*); this area depends upon the operating parameters, but can be several micrometres (*Table 6.4*). Thus, patterns are subject to distortion if there is buckling of the foil or a change in the thickness of the sample. A further disadvantage over the standard method is that a shadow image of the specimen is superimposed on the diffraction pattern. However, there are applications in which a mixture of diffraction and image information can be advantageous (Spence and Zuo, 1992).

Figure 6.11. Bright-field Tanaka pattern from the 24R polysome of the Al, Sn oxide mineral nigerite. [001] zone axis at 150 kV. The pattern was recorded at a camera length of 1050 mm in the microprobe mode of a Philips EM430 using a defocus value of 60 µm and a 200 µm C2 aperture. The circle below the pattern represents the size of the largest CBED disc that could be used without overlap in a conventional CBED pattern at this orientation.

Table 6.5. The Tanaka method: values of the defocus, Δf, that are needed and the diameter, D, of the area of the specimen contributing to the pattern for a range of operating parameters

α mrad (degrees)	p (nm)	Diffraction aperture (µm)	$D/\Delta f$ (µm/µm)			
10 (.57)	4	0.2	*	*	0.1/0.5	0.02/1
20 (1.1)	30	1	*	0.8/1.5	0.16/4	0.32/8
40 (2.3)	250	9	0.5/6.7	1.0/13	2.5/34	5/67
60 (3.4)	900	30	2.8/24	5.6/47	14/120	28/240
80 (4.6)	2000	70	8.4/52	17/105	42/270	84/540
d-spacing of planes of smallest g in diffraction pattern (nm)			0.1	0.2	0.5	1.0

The calculations are for a microscope operating at 100 kV in which the spherical aberration coefficient C_s = 2 mm and the magnification of the image-forming part of the objective lens is 35 ×. From Eades, 1984.

*$\theta > \alpha$; normal CBED can be used.

α = semi-angle of probe; p = diameter of focused probe.

Exercises

6.1 *Figure 6.12a* is a [001] CBED pattern from a polytype of SiC. Measure the radius of the FOLZ and, from the table below, decide to which polytype the pattern corresponds (hexagonal axes have been used). Voltage 200 kV, $\lambda = 2.51$ pm, camera constant $\lambda L = 1.17$ nm mm.

Polytype	c (nm)	Lattice type
4H	1.0053	Hexagonal P
6H	1.508	Hexagonal P
5R	3.770	Trigonal R
21R	5.278	Trigonal R
27R	6.7996	Trigonal R

Figure 6.12b is an enlargement of the bright-field spot. From the symmetry shown in (a) and (b) determine the possible diffraction groups and point groups. Reproduced with the permission of JEOL Ltd from Tanaka and Terauchi (1985).

6.2 Given that the space group of MgGeO$_3$ is *C2/c* and that the camera constant for *Figure 6.6b* is 0.806 nm mm, determine the lattice repeat along the y-axis. The accelerating voltage was 100 kV ($\lambda = 0.0037$ nm).

(a)

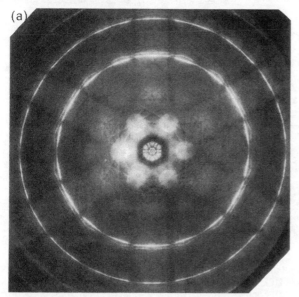

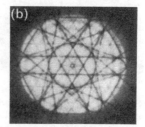

Figure 6.12. (a) [001] CBED pattern from a polytype of SiC. (b) Detail of the bright-field disc. Reproduced from Tanaka and Terauchi, *Convergent Beam Electron Diffraction*, 1985, with permission of JEOL Ltd.

6.3 *Figure 6.13* shows diffraction patterns from the ceramic BaO.Pr$_2$O$_3$.4TiO$_2$. (a) [001] zone-axis pattern, showing reflections in the ZOLZ and FOLZ. (b) Microdiffraction pattern of the [100] zone axis. (c) [010] pattern that has been rotated about the z*-axis. (d) [100] CBED pattern taken at a low camera length and with a high convergence angle. What is the space group of the material?

6.4 Determine the WP (full) symmetry of *Figure 6.2b*. What are the possible diffraction groups and point groups? How could these point groups be distinguished?

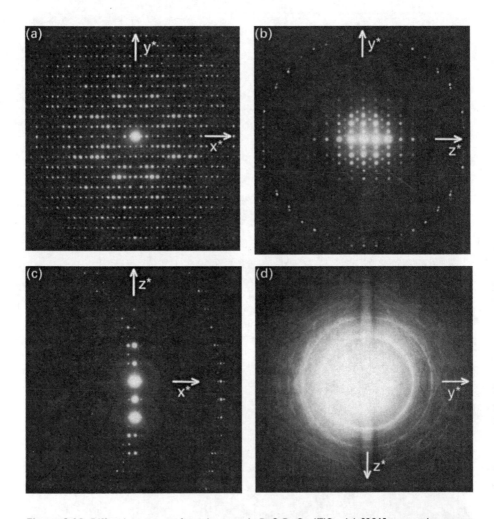

Figure 6.13. Diffraction patterns from the ceramic BaO·Pr$_2$O$_3$·4TiO$_2$. (a) [001] zone-axis pattern, showing reflections in the ZOLZ and FOLZ. (Note: these are much less well separated than usual). (b) Microdiffraction pattern of the [100] zone axis. (c) [010] pattern that has been rotated about the z*-axis to show detail in the FOLZ. (d) [100] CBED pattern taken at a low camera length and with a high convergence angle. Reproduced from Azough F *et al., J. Appl. Cryst.* 1995; 28: 577–581, with the permission of Munskgaard International Publishers Ltd, Copenhagen, Denmark.

6.5 Al$_2$CuLi is known to be hexagonal from X-ray diffraction studies and shows the WP (proj) and WP (full) symmetry given below. What is the point group?

Zone axis	WP (proj)	WP (full)
[001]	6*mm*	6*mm*
[1$\bar{1}$0]	2*mm*	2*mm*
[1$\bar{1}$0]	2*mm*	*m*

6.6 What are the BF (proj) and BF (full) symmetries of the nigerite in *Figure 6.11*? If the WP (full) symmetry is 3*m* and the BF (proj) symmetry is 6*mm*, what is the diffraction group for this pattern and what are the possible point groups?

7 The fine structure in electron diffraction patterns

The fine structure, such as streaks and satellites, present in many electron diffraction patterns contains a great deal of information about the defect structure of the specimen. But, as we shall see, a number of different factors can produce similar effects, and caution must be exercised in interpreting them. In some samples, such as commercial alloys, several different types of defect may be present within a small area. It is therefore imperative to correlate detail in the diffraction pattern with features in the image, in particular by forming dark-field images (Section 3.3.2).

In this chapter, we shall look at:

- extra spots;
- streaks and diffuse reflections;
- spot splitting and satellites;
- walls of intensity.

7.1 Double diffraction

As we have seen in Section 5.3, electrons are very strongly scattered by atoms (several orders of magnitude more strongly than X-rays, for example). As a consequence, a diffracted beam travelling through a crystal can be rediffracted either within the same crystal or when it passes into a second crystal such as a precipitate or twin, as shown in *Figure 7.1a*. The process, known as double diffraction, is shown in terms of the reciprocal lattice and the reflecting sphere in *Figure 7.1b*. The incident beam is AO and, as the reciprocal lattice point $h_1k_1l_1$ lies on the reflecting sphere at B, a diffracted ray CB is produced. The diffracted ray becomes a potential new incident ray EO with its own reflecting sphere centred at C'. A second reciprocal lattice point $h_2k_2l_2$ will lie on the new reflecting sphere at F and a doubly diffracted ray C'F is produced.

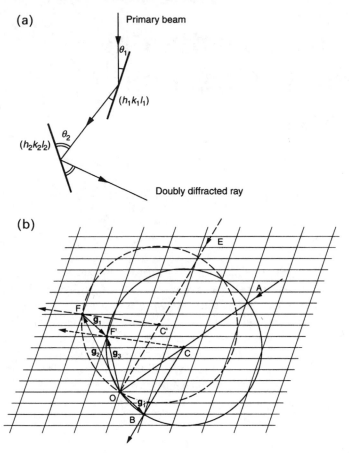

Figure 7.1. Double diffraction. (a) The primary beam is incident on a set of planes $(h_1k_1l_1)$ at their Bragg angle θ_1. The diffracted beam is then rediffracted from a second set of planes $(h_2k_2l_2)$ at their Bragg angle θ_2. (b) The process in terms of the reciprocal lattice and the reflecting sphere. Note that the size of the reflecting sphere relative to the reciprocal lattice is appropriate for X-ray, rather than electron, diffraction. See text for details.

However, the reflection will appear to have come from the set of planes $(h_3k_3l_3)$ whose reciprocal lattice point lies at F'. If the vector OB is $\mathbf{g}_1$, OF is $\mathbf{g}_2$ and OF' is $\mathbf{g}_3$, the indices of the reflection $h_3k_3l_3$, can be found by vector addition. As FF' = OB = $\mathbf{g}_1$:

$$\mathbf{g}_3 = \mathbf{g}_2 + \mathbf{g}_1$$

and:

$$h_3k_3l_3 = h_1 + h_2, \ k_1 + k_2, \ l_1 + l_2$$

Note that the reciprocal lattice point $h_3k_3l_3$ is on the original reflecting sphere and, if it is an 'allowed' reflection (i.e. one with a non-zero structure factor), the double diffraction process will merely add to the intensity produced from the $(h_3k_3l_3)$ planes by the kinematic diffraction process (i.e. conditions under which no double diffraction takes place).

However, if $h_3k_3l_3$ is 'forbidden' (i.e. its structure factor is zero) an extra reflection will appear in the diffraction pattern, as explained below.

7.1.1 *Double diffraction of 'forbidden' reflections*

In Section 2.8.2, we noted that certain reflections that are forbidden for the hexagonal close-packed structure appear in the diffraction pattern by double diffraction. Let us look at the [100] = [2$\bar{1}\bar{1}$0] pattern (see *Figure 2.15a*) as an example. One set of reflections that are 'forbidden' is 000l where l is odd. These reflections can appear in the pattern *via* a number of routes: for instance, because 01$\bar{1}$1 + 0$\bar{1}$10 = 0001, the 0001 reflection can appear by diffraction firstly from the (01$\bar{1}$1) planes and then by rediffraction by the (0$\bar{1}$10) planes (or *vice versa*). A graphic way of seeing how doubly-diffracted spots arise is to make two copies of the pattern, the second on tracing paper, and to superimpose the second on the first without rotation, but with the origin of the second pattern displaced to a spot of the first pattern. If you do this for the example given above and displace the origin of the second pattern to the 01$\bar{1}$1 reflection of the first, you will see that the 0$\bar{1}$10 reflection in the second pattern lies above the position of the forbidden 0001 reflection in the first pattern.

The forbidden reflections described above are the result of the presence of a symmetry element containing a translation (a *c*-glide plane parallel to {1$\bar{1}$00}; see Appendix B for an introduction to space groups). Such 'absent' reflections normally reappear in diffraction patterns by double diffraction, but there is one situation where they cannot do so, at least from interactions within the zero-order Laue zone (ZOLZ). This situation occurs when a glide plane is perpendicular to the electron beam. An example is the [010] zone-axis pattern from a crystal that contains a *c*-glide plane parallel to (010) (*Figure 6.3*). The conditions imposed by the glide plane are that $h0l$ reflections will be absent if l is odd (see *Table 5.1*). As *all* the reflections in the [010] zone-axis pattern have indices $h0l$, half of the reflections will be missing and there is no route for double diffraction to occur from within the zero layer of the reciprocal lattice (try it!). In the [100] zone-axis pattern, on the other hand, the reflections have indices $0kl$ and absences will only occur along the z* direction, i.e. 00l when l is odd, and therefore there is a route for double diffraction to occur. This situation is similar to that of the example of the hexagonal crystal above. Other examples of patterns in which absences that are the result of glide planes show systematic absences that do not reappear by double diffraction from within the ZOLZ are shown in *Figures 6.2* and *6.5*. Note, however, that systematic absences that occur in the ZOLZ as a result of a glide plane perpendicular to the zone axis can reappear by double diffraction from planes whose reflections appear in the first-order Laue zone (FOLZ). An example is the <001> diffraction pattern of spinel (*Figure 6.2*) in which the 'forbidden' 200 and 020 reflections can occur by a route such as 1 21 1 + 1 $\overline{21}$ $\bar{1}$ = 200 (Smith, 1978).

No reflections that are absent due to the lattice type not being primitive can reappear by double diffraction. Try finding two indices of 'allowed' reflections in the [001] pattern from the face-centred pattern in *Figure 2.13c* that add together to give the indices of a forbidden reflection such as 100. You will find that it is not possible because all the reflections in this section have indices that are even numbers. Similarly, double diffraction of reflections in the FOLZ cannot produce forbidden reflections because all the reflections in the FOLZ have indices that are odd numbers.

The conditions for absence due to certain lattice types and translational symmetry elements are derived in Chapter 5 and an introduction to space group symmetry is given in Appendix B.

7.1.2 *Double diffraction of spots from precipitates and twins*

Double diffraction can also occur between reflections from different phases and twins, and quite complex patterns can result. *Figure 7.2a* shows a diffraction pattern from an Al–Mg–Si alloy containing Mg_2Si precipitates in which the double diffraction effects are particularly striking. Both phases are face-centred cubic and they are oriented with the relationship:

$$(100)_{matrix}//(110)_{precipitate}$$

$$[001]_{matrix}//[001]_{precipitate}$$

but the spacing of the (110) planes of the precipitate is slightly larger than the spacing of the (100) planes of the matrix so none of the precipitate and matrix spots coincide in the diffraction pattern. The aluminium matrix pattern and the primary pattern from Mg_2Si are shown indexed in *Figure 7.2b*. If the matrix reflection 020 is allowed to act as a secondary source for diffraction in the precipitate, the spots shown as open circles in *Figure 7.2c* result. When *all* the aluminium spots are considered as possible secondary sources in this way, the pattern in *Figure 7.2d* is obtained. You can verify this by using the overlay technique described above.

The multiple grouping of spots resulting from double diffraction is very common and can also result from surface films and twins (*Figure 7.3*).

7.1.3 *Detection of double diffraction in the microscope*

You may want to know if particular diffraction spots arise from double diffraction. As the process will only occur if dynamical conditions prevail (i.e. when multiple scattering events can take place) the specimen has to have a certain minimum thickness for double diffraction to be seen in the diffraction pattern. This thickness depends on the structure of the specimen, but for a typical silicate, for example, would be about 50 nm. It follows that if the specimen is sufficiently thin, you will be able to record a

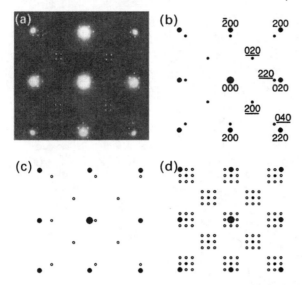

Figure 7.2. (a) Diffraction pattern from a Mg$_2$Si precipitate in an aluminium matrix. (b)–(d) The interpretation of (a). The large spots represent aluminium reflections and small spots represent precipitate reflections; open circles represent spots arising from double diffraction. (b) Primary patterns. (c) Double diffraction arising from the 020 aluminium reflection as a secondary source. (d) Complete pattern resulting from double diffraction from the other Al spots acting as secondary sources, giving little squares of doubly-diffracted spots, arising in effect from the big square of the Al pattern. Adapted from Hirsch PB *et al.*, *Electron Microscopy of Thin Crystals*, 1965, with the permission of Butterworth Heinemann, London.

diffraction pattern under kinematical conditions and the systematic absences will be apparent, as in *Figure 6.5a*.

Since double diffraction relies on two diffracted beams being excited at once, it is useful to tilt the specimen about 10° from the zone axis about the direction containing the reflection that you suspect is doubly diffracted so that the reflections responsible for the double diffraction are no longer on the reflecting sphere. If the suspect spot is doubly diffracted, it will first be strongly reduced in intensity and eventually disappear. An example is shown in *Figure 7.4*. Here, the $0k0$ reflections with k odd disappear when the crystal is rotated about y* so that only a systematic row of $0k0$ reflections is present in the ZOLZ. This is because, for example, the 'forbidden' 010 reflection requires a route such as $011 + 00\bar{1} = 010$ to appear in the diffraction pattern and the (011) and $(00\bar{1})$ planes are no longer diffracting in *Figure 7.4b*. Note that if diffraction is limited to a row of systematic reflections, double diffraction can still occur between the excited reflections. An example of this is *Figure 7.2*; if the pattern is tilted about x* or y* the row of extra spots parallel to the axis would remain.

It is also possible to detect reflections that arise from double diffraction by dark-field imaging (see Section 3.3.2). If the precipitate particle or twin intersects the specimen surfaces, the double-diffraction process will be

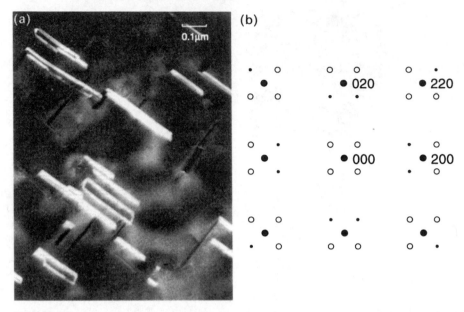

Figure 7.3. (a) {111} twins in a (001) gold film. The dark-field image was formed using a beam produced by double diffraction at the boundary between the twins and the matrix; consequently, only the boundaries appear bright. Reproduced from Pashley DW and Stowell MJ, *Phil. Mag.* 1963; 8: 1605–1632, with the permission of Taylor and Francis. (b) Schematic [001] diffraction pattern for the foil in (a). The large spots represent matrix reflections and small spots represent primary twin reflections; open circles represent spots arising from double diffraction.

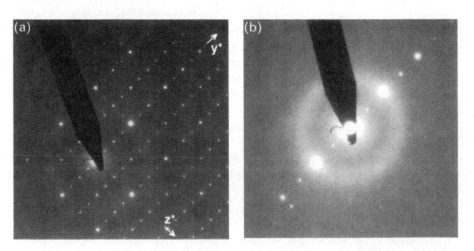

Figure 7.4. Diffraction patterns from the synthetic leucite $Cs_2CuSi_5O_{12}$, which is tetragonal. (a) Approximately [100] zone-axis pattern. (b) The crystal has been rotated about the y^*-axis so that only the $0k0$ systematic row of reflections is strongly excited. The $0k0$ reflections with k odd have disappeared because they are forbidden by virtue of the screw diad parallel to the y-axis in the space group, $P4_2/n\ 2_1/c\ 2/m$. (see *Table 5.1*). The 010 reflection, for instance, requires a route such as 011 + 00$\bar{1}$ = 010 to appear in the diffraction pattern and the (011) and (00$\bar{1}$) planes are no longer diffracting in (b).

confined to the interface, which will be bright in the dark-field image using a doubly diffracted spot (*Figure 7.3*). However, if the particle is entirely confined within the specimen, there will be no obvious differences between dark-field images with doubly diffracted or primary precipitate spots.

7.2 Diffraction from small particles

In Section 3.1, we saw that the thin dimension of specimens used in the TEM results in a streak or **relrod** through each node of the reciprocal lattice. The relrod is perpendicular to the thin dimension of the crystal and its length is $1/t$, where t is the thickness of the crystal. This **shape effect** applies equally to small particles whose atomic constituents have a different scattering power, f, from that of the matrix. The shape of the reciprocal lattice points or **relps** depends on the shape of the particles. The shape factors are shown for various shapes of particles in *Figure 7.5*. The point to remember is that, in the reciprocal lattice, 'small become large' and *vice versa*: if the particle is a cube with side of length A, in the diffraction pattern there is a rod of diffuse intensity of length approximately $1/A$ perpendicular to each of the faces of the cube; if it is spherical, there is a diffuse shell around each relp; if it is disc shaped, there is a streak through each relp perpendicular to the plane of the disc;

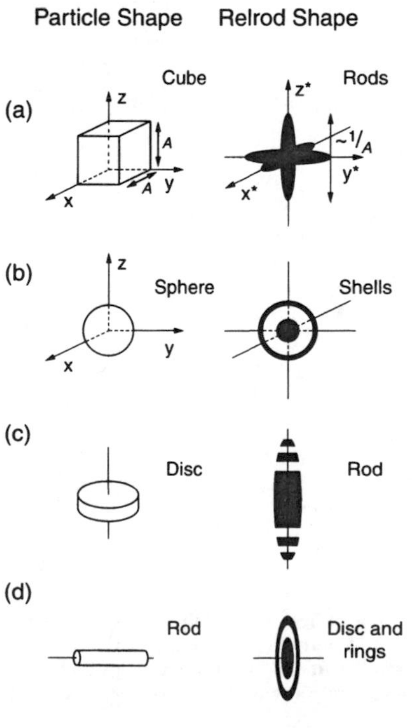

Particle Shape   Relrod Shape

Figure 7.5. The shapes of reciprocal lattice 'points', or relps, for various shapes of particles. Note that the subsidiary maxima shown are so weak that they are unlikely to be visible in a diffraction pattern.

and if it is rod shaped, there are sheets of intensity through each relp. One example of disc-shaped particles is the presence of the G-P zones in *Figure 3.10*. The (100) and (010) platelets are oriented parallel to the beam and streaks are visible in the diffraction pattern along the x* and y* directions. The habit plane of the platelets can be determined because it is perpendicular to the direction of the streaks.

7.2.1 *The distribution of streaks in the diffraction pattern*

Because the diffraction pattern is a section through the three-dimensional reciprocal lattice, streaks in the reciprocal lattice can produce either streaks or satellites in the diffraction pattern. If $s = 0$, a streak tangential to the reflecting sphere will be visible in the diffraction pattern (*Figure 7.6a*). If, however, $s \neq 0$, the reflecting sphere cuts the streaks at an angle such that one or more satellites are formed in the diffraction pattern (*Figure 7.6b*). A satellite pair formed in this way can easily be

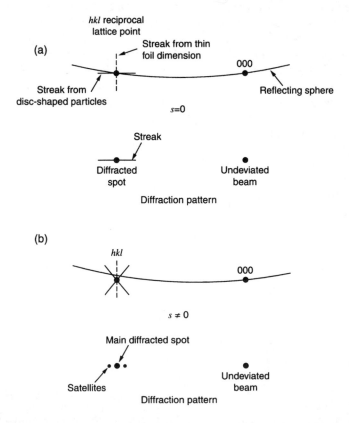

Figure 7.6. (a) Streaking of reciprocal lattice points arising from disc-shaped particles lying parallel to the electron beam, $s = 0$. (b) Satellites about a reciprocal lattice point resulting from the intersection of the reflecting sphere with inclined streaks at $s \neq 0$. Adapted from Edington JW, *Electron Diffraction in the Electron Microscope*, 1975, with the permission of the Macmillan Press.

distinguished from those resulting from spinodal decomposition or periodic antiphase domain boundaries (see Sections 7.4.1 and 7.4.2) because, in the latter case, the spacing of the satellites from the matrix spot is constant whereas, in the former case, the spacing depends on the value of s.

For the discs of intensity produced by rod-shaped particles (*Figure 7.5d*), the streaks in the diffraction pattern will be straight if the length of the rods is perpendicular to the incident beam direction. However, if the rods are at some other angle to the incident beam, the streaks will be curved because of the curvature of the reflecting sphere. It is clear that if the shape of particles is to be investigated from the analysis of the streaks that they produce in the diffraction pattern, it is important to tilt the sample carefully.

7.2.2 The effects of elastic strain

We have so far only considered the effect on the diffraction pattern of the size and shape of particles that have a different scattering power from the matrix. However, **coherent** particles (those whose lattice planes are continuous with those of the matrix) produce an elastic distortion of the matrix if the cell parameters of the two phases are different. This distortion produces diffuse scattering in reciprocal space in the direction of the distortion. The effect can be quite intense, and the diffraction pattern can be a more sensitive indication of the presence of small, coherent particles than the image. However, distortional effects are often present at the same time as shape effects and the two are difficult to separate because they usually occur in the same direction (e.g. in Al–Cu, *Figure 3.10*).

The main characteristics of uniaxial elastic strain effects are described by Guinier (1958) and are listed as follows (*Figure 7.7*).

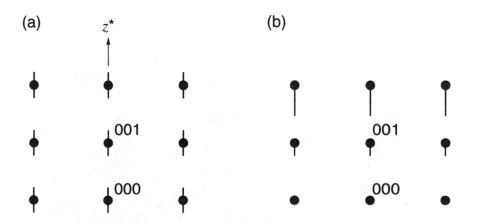

Figure 7.7. Schematic diffraction patterns for different types of particles. (a) Thin, strain-free disc on (001) in which there is a difference in scattering amplitude, f, between the matrix and particles. (b) Distorted, coherent disc on (001) where the unit-cell volume of the particle is smaller than that of the matrix. There is no difference in the scattering amplitude, f, between the particle and the precipitate.

- The length of the streak increases with increasing **g** (although this is difficult to observe as each reflection must be tilted to $s = 0$).
- There are no streaks through the undeviated beam.
- There are no streaks at spots that arise from planes parallel to the direction of the distortion, e.g. a distortion along [001] will cause streaking of the 001 spot, but not of the 100 or 010 spots.

In principle, the above characteristics afford a method of distinguishing between shape and distortion effects, but the occurrence of multiple scattering makes this difficult, except in very thin crystals.

7.3 Diffraction from ordered structures

7.3.1 Long-range order

At high temperatures, many metal alloys, minerals and ceramics have disordered structures, i.e. some or all of the atoms are randomly distributed on the atomic sites. At lower temperatures, the atoms may show a preference for certain sites and the structure will become ordered. An example is Al_3Li, which is face-centred cubic at high temperatures, with the Al and Li randomly occupying atomic positions at the lattice points. But at lower temperatures the Li atoms occupy the corner positions while the aluminium atoms occupy the face-centring positions; as a consequence the lattice becomes primitive (*Figure 7.8a*). The result is that **superlattice reflections** that are forbidden for the face-centred lattice (those for which h, k and l are neither all odd nor all even) appear in the diffraction pattern of the ordered phase (*Figure 7.8b*). The superlattice reflections are always less intense than the main reflections

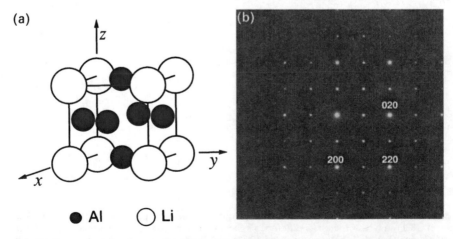

(a)   (b)

● Al   ○ Li

Figure 7.8. (a) The structure of Al_3Li. (b) [001] diffraction pattern of Al_3Li. Diffraction pattern reproduced by courtesy of P.P. Prangnell.

because their intensity depends upon the difference between the scattering factors of the ordering atoms (see Section 5.2.2).

Another example of an ordered structure of this type is CuZn which is body-centred cubic at high temperatures and orders to the CsCl structure (Section 5.2.2) at low temperatures, with the Cu (or Zn) taking up the body-centring position. In this case, the weak superlattice reflections are those for which $h + k + l$ is odd.

In other superlattice structures, the unit cell increases in size as a result of atomic ordering. An example is shown in *Figure 7.9b* of a diffraction pattern from the mineral nigerite, a tin-and titanium-rich oxide of aluminium and zinc. The 'parent' structure, gahnite, $ZnAl_2O_4$, is face-centred cubic (*Figure 7.9a*), but, in nigerite, the ordering of the tin and titanium in layers parallel to the (111) planes of the parent structure results in the unit cell becoming four times larger in that direction (so the reciprocal lattice repeat is one quarter of that of the parent structure). The unit cell of nigerite is trigonal, rhombohedral, and the superlattice direction is the z-axis, $c = 1.86$ nm. There are a whole series of **polysomes** of nigerite with different z-axis repeats, the value of which depend upon the tin plus titanium content of the unit cell.

7.3.2 Short-range order

Point defects are defects that occur on atomic sites. They may be **substitutional** atoms, vacancies (empty atomic sites) or **interstitial**

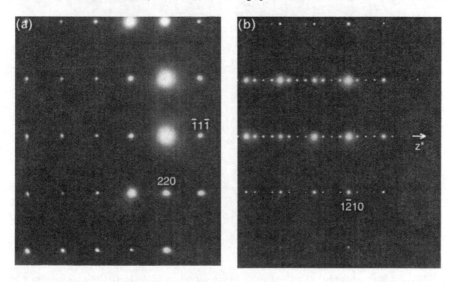

Figure 7.9. (a) The [$\bar{1}12$] diffraction pattern from gahnite, $ZnAl_2O_4$, which is face-centred cubic. (b) [210] diffraction pattern from nigerite, a tin- and titanium-rich oxide of aluminium and zinc. In the nigerite, ordering of the tin and titanium in layers parallel to ($\bar{1}1\bar{1}$) of the 'parent' gahnite structure results in the repeat along [$\bar{1}1\bar{1}$]$_{fcc}$ becoming four times larger (1.86 nm) and the unit cell becoming trigonal, rhombohedral. Note that [$\bar{1}1\bar{1}$] of the gahnite becomes z* in nigerite because of the change in the choice of unit cell.

atoms (ones that takes up positions between normal atomic sites). These point defects can order to give a superlattice, as described above, but if there is an insufficient number to give long-range order, **short-range order** can occur. This short-range ordering can give rise to diffuse scattering and has been described in non-stoichiometric metal carbides with composition around M_6C_5, in which the carbon vacancies have ordered in octahedral sites around the metal atoms. *Figure 7.10* shows three examples of diffraction patterns from vanadium carbide, which is face-centred cubic. The diffuse intensity occurs as circles and arcs, sometimes around spots and sometimes not. Sauvage and Parté (1972) showed that the diffuse intensity forms a surface in reciprocal space as shown in *Figure 7.10d*; the diffuse scattering in diffraction patterns is explained by the intersection of the reflecting sphere with this surface.

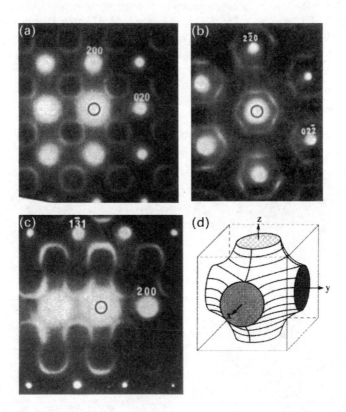

Figure 7.10. The effect of short-range ordering of vacancies on the (a) [001], (b) [111] and (c) [013] diffraction patterns of non-stoichiometric vanadium carbide. (d) The surface formed by the diffuse intensity in reciprocal space. (a)–(c) reproduced from Lewis MH and Billingham J, *JEOL News*, 1972, with the permission of JEOL USA Inc.; (d) modified from Sauvage and Parté (1972).

7.4 Periodic and modulated structures

Periodic variations in interplanar spacings, variations in structure factor (i.e. atomic species; see Section 5.2), or both together; or a periodic distribution of crystal defects such as antiphase domain boundaries, crystallographic shear planes or dislocations can lead to the occurrence of satellites in the electron diffraction pattern.

7.4.1 Spinodal decomposition and other compositional modulations

Spinodal decomposition is a mechanism by which a supersaturated solid–solution may undergo phase separation without a nucleation stage; instead there is a gradual build-up of composition modulations, which is sinusoidal in the initial stages, in the elastically soft directions in the crystal (see, for instance, Porter and Easterling, 1992). A pair of satellite reflections is present in the reciprocal lattice on either side of each diffracted spot at a distance of $1/\lambda$ from it, where λ is the wavelength of the modulation. The separation of the satellite spots is in the direction of the modulation (*Figure 7.11*).

Any regular intergrowths of two phases or compositional modulations will produce satellite reflections in the diffraction pattern. *Figure 7.12* shows a regular intergrowth of magnetite, Fe_3O_4, and amorphous silica, SiO_2, produced by heating the natural mineral fayalite, Fe_2SiO_4, in air. The lamellae of magnetite (the phase showing the dark contrast) have a regular spacing of about 15 nm and are parallel to (110). In the diffraction pattern each spot is surrounded by a row of satellites, spacing 15 nm^{-1},

Figure 7.11. Electron micrograph of a natural alkali feldspar, $Na_{0.64}K_{0.36}AlSi_3O_8$, that has undergone spinodal decomposition to produce composition modulations parallel to approximately ($\overline{6}01$). Inset is an enlargement of one of the diffraction spots that shows satellites in the direction of the modulations. The image and the diffraction pattern are in their correct relative orientation. Reproduced from Owen DC and McConnell, *Nature Phys. Sci.*, 1971; 230: 118–120 with the permission of Macmillan Magazines Limited, London.

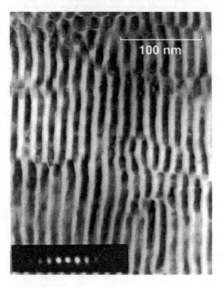

Figure 7.12. Electron micrograph of a regular intergrowth of magnetite, Fe_3O_4, and amorphous silica, SiO_2, produced by heating the natural mineral fayalite, Fe_2SiO_4, in air for 3 days at 800°C. Inset is an enlargement of the undeviated beam, showing satellites parallel to g_{110} of the magnetite. The image and diffraction pattern are in the correct relative orientation.

parallel to g_{110} (inset, *Figure 7.12*). Note that here there is a row of satellites, rather than a single pair, as occurs with spinodal decomposition, because the modulation is a square wave rather than a sinusoidal one (the explanation lies in the mathematical device known as the **Fourier Transform** of the two wave forms).

7.4.2 Periodic antiphase domain boundaries

At high temperatures, the atoms in the structure of CuAu are disordered and the lattice is face-centred cubic, but at lower temperatures the atoms order so that there are alternate layers of gold and copper atoms parallel to the (100) planes and the lattice becomes primitive (*Figure 7.13a*). This kind of ordering gives rise to **antiphase domains**: domains in the structure that are 'out-of-step' with each other by half a lattice repeat because the ordering can begin on one of two atomic positions that are equivalent in the disordered structure, i.e. the 'corner' or 'face-centring' positions, but which become distinct in the ordered structure (*Figure 7.13a*). An **antiphase domain boundary**, or **APB**, exists between regions ordered on different 'registers', i.e. a change-over from, say, Au occupying the corner positions to their occupying the face-centring positions. In CuAuII, the **antiphase vector** (the vector that would bring the two regions back into register) is 1/2[110], i.e. one half of the diagonal of the face of the unit cell. In CuAu, the antiphase domains are periodic with a domain boundary, on average, every five unit cells along the x-axis (*Figure 7.13a*). As a result of the periodicity of the APBs, diffraction patterns contain satellite spots (*Figure 7.13b*). *Figure 7.13c* shows the row of satellites produced by the periodic APBs from one orientation of the array (there are three possible orientations, all of which will occur,

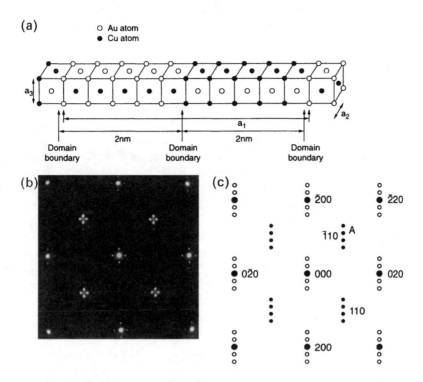

Figure 7.13. (a) The unit cell of the periodic antiphase structure in AuCuII. (b) [100] electron diffraction pattern from AuCuII. (c) The position of the satellites produced by one orientation of the periodic APBs. The small, closed circles are the primary satellites and the small, open circles are mainly produced by double diffraction. Notice that there is no intensity at the positions of the superlattice reflections such as 110, only satellites on either side of the positions. Adapted from Pashley DW and Powell MJS, *J. Inst. Metals*, 1959; 87: 419–428, with the permission of the Institute of Materials.

because the x-, y-and z-axes are symmetrically equivalent in the cubic system). The primary satellites occur around the superlattice positions such as 110 (indexed on the original face-centred cubic cell) and their distance from the superlattice positions is $1/\lambda$, where λ is the periodicity of the APBs. Double diffraction largely accounts for the satellites on the undeviated beam and the main spots, although there is a small contribution to these satellites from the periodicity of the structure that can be ascribed to the small change in lattice parameter that occurs at the APB, i.e. the effect described in Section 7.2.2 (Edington, 1975).

7.4.3 Regular dislocation arrays

Regular dislocation arrays, such as occur at interphase boundaries, can produce weak satellites perpendicular to the dislocation lines. Examples are shown by Edington (1975) and Williams and Carter (1996). Note that, because of the likelihood of double diffraction, careful tilting to two-beam conditions (Section 4.1) is essential to eliminate double diffraction and be certain of the origin of the satellites.

7.4.4 Incommensurate structures

In modulated structures and superlattices that involve an increase in the size of the unit cell, the wavelength of the fluctuations is often a simple multiple of the unit cell of the basic substructure, in which case the modulation is said to be **commensurate**. Examples are the periodic antiphase domains in *Figure 7.13*, in which domain boundaries occur, on average, every five basic (disordered) unit cells, and the four-fold superlattice in *Figure 7.9*. However, if the periodicity of the fluctuations is not rationally related to the spacing and/or the orientation of the basic periodicity, the modulations are said to be **incommensurate**. In the Cu–Au system, deviation from CuAu stoichiometry, or the addition of small quantities of other elements, causes the average periodicity of the APBs to vary and the structure becomes incommensurate. The distance of the satellite spots from the superlattice positions in the diffraction pattern is then an irrational fraction of the g_{100} spacing.

The term 'incommensurate' can also be applied to structures such as the spinodal intergrowth in *Figure 7.11* or the magnetite/silica intergrowth in *Figure 7.12*.

7.5 Planar and linear defects

Stacking faults and thin twins, like small disc-shaped particles, will cause streaking in the electron diffraction pattern (see *Figure 3.9a*). If a set of parallel lattice planes in the crystal is arranged irregularly, either because the distance between them varies or because they are displaced randomly parallel to themselves by a non-constant amount, then the scattering is limited to rows of the reciprocal lattice that are normal to the lattice planes (*Figure 7.14*).

If, instead of the lattice planes being displaced as described above, the planes are regularly spaced, but displaced randomly parallel to themselves by a specific fraction of the unit-cell dimensions, some of the diffraction spots will be sharp and others will be diffuse. A simple example of such stacking disorder, as shown by the mineral wollastonite, $CaSiO_3$, is illustrated in *Figure 7.15*. Layers of the structure parallel to (100) are randomly displaced a distance $b/2$. The corresponding diffraction pattern is shown schematically in *Figure 7.15b*; some spots are sharp while others are diffuse and elongated in a direction normal to the fault planes. The diffuse reflections are diffuse only in the x^* direction because the atomic arrangement in the individual (100) planes is perfect. Because the relative displacement of neighbouring (100) planes is $b/2$, the diffuse

(a)

Direct lattice

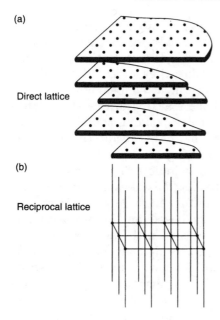

(b)

Reciprocal lattice

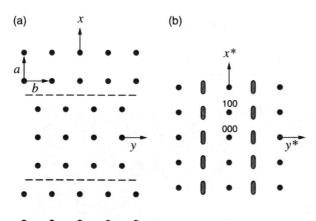

Figure 7.14. (a) Planar disorder in which the planes are identical and parallel, but arranged irregularly. (b) Scattering in reciprocal space is limited to rows normal to the planes in the direct lattice. Reproduced from Guiner A, *X-ray Diffraction in Crystals, Imperfect Crystals and Amorphous Bodies*, 1963, with the permission of W.H. Freeman and Co.

(a)   (b)

(c)

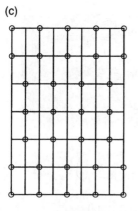

Figure 7.15. (a) Diagram representing the stacking shown by the mineral wollastonite, $CaSiO_3$, projected onto the (001) plane. The dots represent lattice points, which, in turn, represent six $CaSiO_3$ 'units'. (b) Schematic diagram of the [001] diffraction pattern corresponding to (a). (c) Diagram of the stacking lattice which corresponds to all the lattice points of (a). Reproduced from Willis BMT, *Proc. Roy. Soc. Lond.*, 1958; A248: 183–198, with the permission of the Royal Society.

spots are those for which $k \neq 2n$, where n is an integer. The reason for this can be understood by reference to the **stacking lattice**, which is represented by the full lines in *Figure 7.15c*. All the lattice points in *Figure 7.15c* lie on this lattice, and therefore they must all scatter in phase in directions corresponding to the diffracted beams from this lattice. Since the horizontal cell edge of the stacking lattice is *half* the length of the lattice of *Figure 7.15a*, the corresponding cell edge of the reciprocal lattice must be *twice* as long as that of the reciprocal lattice of *Figure 7.15b*. Therefore, the sharp spots of *Figure 7.14b* correspond to diffraction from the stacking lattice and have indices for which $k = 2n$.

The probability of occurrence of the stacking mistakes determines the length and distribution of intensity of the diffuse spots. *Figure 7.16* shows four optical diffraction patterns of masks based on the wollastonite structure in which the probability of a mistake in the stacking, p, has been varied. The sharp spots are unchanged throughout and correspond to the common stacking lattice. For a small fault with a probability such as $p = 0.1$ (i.e. on average one in 10 planes will be displaced) the streaking of the $k \neq 2n$ spots is not very marked, but for $p = 0.3$ it is significant. For $p = 0.8$, the majority of the structure is face centred and the diffraction pattern shows diffuse reflections in the face-centring positions. When $p = 0.5$, there is equal probability of either arrangement and the diffuse spots extend continuously along x^*.

Finally, if identical, parallel lattice rows of spacing a in the crystal are arranged irregularly with respect to each other, the reciprocal lattice

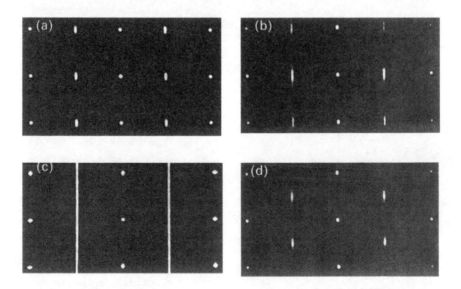

Figure 7.16. Four optical diffraction patterns corresponding to masks based on the faulting in wollastonite, $CaSiO_3$ (*Figure 7.15a*) with probabilities (a) $p = 0.1$, (b) $p = 0.3$, (c) $p = 0.5$, (d) $p = 0.8$. Reproduced from Willis BMT, *Proc. Roy. Soc. Lond.*, 1958; A248: 183–198, with the permission of the Royal Society.

consists of a family of planes, normal to the rows, and spaced at a distance of $1/a$.

7.6 Thermal diffuse scattering

Walls or sheets of diffuse intensity ascribed to directional thermal vibrations have been reported in a number of metals. These walls lie on {110} reciprocal lattice planes for Al, Au, Si and Ge, and on {111} for α-Fe, and manifest themselves as diffuse streaks in well-focused (C2 completely overfocused) diffraction patterns taken with exposures of 1–2 min (*Figure 7.17*). The walls are the result of the scattering of rows of atoms in the <110> and <111> directions, respectively; these are the directions joining nearest-neighbour atoms in the two structures. It is supposed that a prominent mode of thermal vibrations is one in which a whole row of atoms along a <110> and <111> direction, respectively, is displaced as a unit along its length. The diffuse scattering effects have the following characteristics:

- their intensity decreases with decreasing temperature and is particularly strong near the melting point;
- the patterns do not change greatly if the specimen is tilted by up to about 10° from the zone axis.

Exercises

7.1 Show that the 'absences' due to a screw diad parallel to the y-axis (Exercise 5.1) will reappear by double diffraction in electron diffraction patterns from crystals of normal thickness.

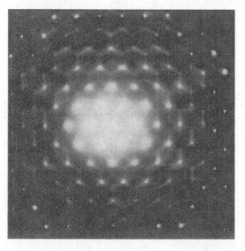

Figure 7.17. The effect of thermal diffuse scattering on the [111] diffraction pattern of silicon. Reproduced from Rymer TB, *Electron Diffraction*, 1970, with the permission of Methuen, London.

7.2 *Figure 7.18* is a [100] diffraction pattern from a disordered silicate. On which planes are the faults and what is the fault vector? Estimate the probability of a fault occurring.

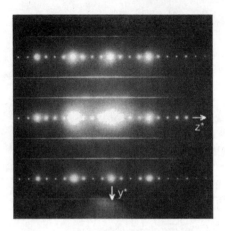

Figure 7.18. [100] diffraction pattern from a disordered silicate.

Appendix A

Some basic crystallography

If you are to study crystalline materials in the electron microscope you need to understand the basic principles of crystallography as set out below. For more details, see one of the texts listed in Section A.6.

A.1 Lattices, crystal systems and crystal classes

Crystals are regular, three-dimensional arrangements of atoms or molecules. These regular arrays or **crystal structures** give rise to internal and external symmetry. The repetitive nature of the arrangements of atoms or molecules in crystal structures leads to the concept of the **crystal lattice** (an array of atoms or molecules at points in space with identical surroundings). There are 14 **Bravais lattices**, i.e. ways of arranging repeating points in space (*Figure A.1*). Each lattice point represents a group of atoms, or motif, with identical surroundings. These lattices amount to seven distinct parallelepipeds or **unit cells** that are primitive (sometimes called simple), symbol P (or R, signifying rhombohedral), i.e. they have lattice points only at the corners of the unit cell, and seven other unit cells that have the same shape as one of the primitive ones, but with more than one lattice point in the cell. These are **centred cells**. The centring may be at the centre of the cell, called **body centring**, symbol I, at the centre of one face of the cell, usually taken to be the x–y face of the cell, symbol C (sometimes called **base centred**), or at the centres of all the faces of the cell, symbol F (**all-face centring**).

Crystals are classified on the basis of their external or morphological symmetry into seven **crystal systems** (**triclinic, monoclinic, orthorhombic, tetragonal, hexagonal, trigonal** and **cubic** (sometimes called **isometric**)) (*Table A.1*). The seven primitive unit cells of the Bravais lattices correspond to five of the seven crystal systems. The exceptions are the hexagonal and trigonal systems; trigonal crystals (defined as having one three-fold axis of symmetry) may have a primitive hexagonal lattice or a rhombohedral lattice. This concept is a rather difficult one and need

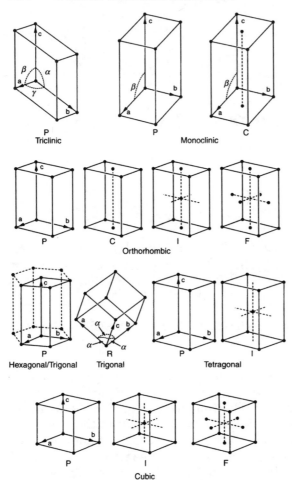

Figure A.1. The unit cells of the 14 Bravais lattices.

only concern you if you are working with trigonal crystals. (For further details see Hahn, 1984 or Hammond, 2001.)

The classification of the crystal systems is based on a **characteristic** or **essential** symmetry (*Table A.1*). Crystals may have more than the essential symmetry and a further subdivision may be made into 32 **crystal classes** or **crystallographic point groups**. These are listed in *Table A.1* in terms of the standard **Hermann–Mauguin** symbols used. For an explanation of the meaning of the symbols, see Hahn (1984), Hammond (2001) or Glusker *et al.* (1994).

A.2 Indexing planes

Planes in any of the Bravais lattices can be indexed using a system known as **Miller indices**. First of all, a unit cell is defined (*Figure A.2a*); the unit cells for the different Bravais lattices are shown in *Figure A.1*.

Table A.1. Crystallographic axial and angular relationships and characteristic symmetry of the crystal systems

Crystal classes[1] (point groups)	System	Axial and angular relationships	Characteristic symmetry
1, $\bar{1}$	Triclinic	$a \neq b \neq c$; $\alpha \neq \beta \neq \gamma$	One-fold axis (identity or inversion, i.e. a centre of symmetry)
2, m, 2/m	Monoclinic	$a \neq b \neq c$; $\alpha = \gamma = 90°$; $\beta \neq 90°$	Two-fold axis (rotation or inversion, i.e. perp. mirror plane) this being taken as the y-axis
222, $mm2$, mmm	Orthorhombic	$a \neq b \neq c$; $\alpha = \beta = \gamma = 90°$	Two-fold axes (rotation or inversion) in three perpendicular directions (the x-, y- and z-axes)
4, $\bar{4}$, 4/m, 422, 4mm, $\bar{4}2m$, 4/mmm	Tetragonal	$a = b \neq c$; $\alpha = \beta = \gamma = 90°$	Four-fold axis (rotation or inversion) along the z-axis
$2\bar{3}$, $m3$, 432, $\bar{4}3m$, $m\bar{3}m$	Cubic	$a = b = c$; $\alpha = \beta = \gamma = 90°$	Four three-fold axes each inclined at 54.73° to the x-, y- and z-axes
3, $\bar{3}$, 32, 3m, $\bar{3}m$	Trigonal	(Hexagonal axes)[2] $a = b \neq c$; $\alpha = \beta = 90°$; $\gamma = 120°$	Three-fold axis (rotation or inversion) along the z-axis
		(Rhombohedral axes) $a = b = c$; $\alpha = \beta = \gamma \neq 90°$	Three-fold axis (rotation or inversion) along [111], i.e. the axis that bisects x, y and z
6, $\bar{6}$, 6/m, 622, 6mm, $\bar{6}m2$, 6/mmm	Hexagonal[2]	$a = b \neq c$; $\alpha = \beta = 90°$; $\gamma = 120°$	Six-fold axis (rotation or inversion) along the z-axis

Note: an *n*-fold rotation axis rotates a crystal face 360/*n*° to produce an identical face. An inversion axis, $\bar{n}$, also involves inversion of the face, i.e. a point at x, y, z is inverted to $-x$, $-y$, $-z$.
[1]For an explanation of these symbols, see Hahn (1984) or Hammond (2001).
[2]For the hexagonal and trigonal systems it is customary to take three axes, x-, y- and u-, at 120° to each other and normal to the z-axis.

Axes x, y and z are chosen at angles α, β and γ with unit vectors **a**, **b** and **c**, and magnitudes a, b and c. In electron diffraction we are concerned with sets of crystallographically equivalent planes because it is these that 'reflect' the electrons. The Miller index of a set of planes is defined in terms of the intercept that one of the two planes on either side of the origin makes on the axes of the unit cell; for instance, in *Figure A.2b* the plane nearest the origin cuts the axes at a/h, b/k and c/l, and (hkl) are the Miller indices of the set. The Miller indices are therefore the reciprocals of the fractional intercepts of the plane on the unit-cell axes. *Figure A.2c* shows two planes of the set (221); the plane nearest the origin, O, makes intercepts of $a/2$, $b/2$ and c on the x-, y- and z-axes, respectively. A negative intercept results in a negative component of the Miller indices, written as $\bar{h}$. Note that if we had chosen to label the plane on the other side of the origin we would have derived the Miller indices ($\bar{h}\bar{k}\bar{l}$).

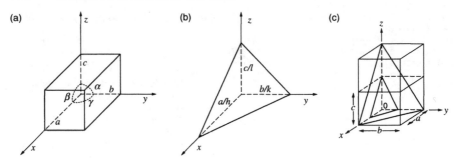

Figure A.2. (a) A unit cell with right-handed axes x, y, z, axial repeats *a, b* and *c* and axial angles *α, β, γ*. (b) The lattice plane from the set (*hkl*) that is nearest to the origin makes intercepts on the x-, y- and z-axes of *a/h, b/k* and *c/l* respectively. (c) Diagram showing two unit cells and the two planes of the set (221) that are closest to the origin, O. The one closest to the origin makes intercepts of *a/2, b/2* and *c* on the x-, y- and z-axes, respectively.

A set of planes that are related by symmetry are known as a form. For instance, in a tetragonal crystal that has a four-fold rotation axis the planes (100), (010), ($\bar{1}$00) and (0$\bar{1}$0) are related by the symmetry. These faces can conveniently be described as the form {100} or {010} etc. Note the use of *curly* brackets here.

Formulae for the angles between different planes in the seven crystal systems are given in Section C.2.

A.2.1 The hexagonal and trigonal systems

For most purposes, the hexagonal unit cell is used for the hexagonal and trigonal systems (*Table A.1*), even though this results in a unit cell three times the primitive one if the lattice is rhombohedral (see Hammond, 2001). In order to emphasize the three- or six-fold symmetry along the z-axis of the unit cell, four indices, known as **Miller–Bravais symbols** (*hkil*) are normally used to index planes in these systems. The extra crystallographic axis, u, is at 120° to the x- and y-axes and normal to the z-axis. As the u-axis is not essential to describe the inclination of a plane in three dimensions, there must be a relationship between *h, k* and *i*. It is: $h + k = -i$.

A.3 Indexing lattice directions

In order to describe a direction in the lattice, we make use of the concept of vectors. In *Figure A.3* the direction OA has the indices [121], i.e. to get from O to A we move one unit along the x-axis, two units along the y-axis and one unit along the z-axis, or in vector notation $\mathbf{r}_{121} = \mathbf{a} + 2\mathbf{b} + \mathbf{c}$. Note the use of *square* brackets for directions. The [110], [$\bar{1}\bar{1}$0] and [01$\bar{1}$] directions are also shown. Note that we can use any origin for the vector; different ones are chosen for convenience for [$\bar{1}\bar{1}$0] and [01$\bar{1}$]. In general:

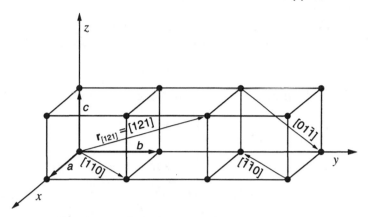

Figure A.3. Directions in the lattice. OA is r_{121} = [121]. Note that we can use any origin for the vector; different ones are chosen for convenience for [$\bar{1}$10] and [01$\bar{1}$].

$r_{UVW} = [UVW] = U\mathbf{a} + V\mathbf{b} + W\mathbf{c}$. We are not concerned with the magnitude of the vector used to describe the direction, so we use the simplest set of whole numbers, i.e. [110] rather than [220] or [½½0].

A set of directions that are related by symmetry have the symbol $<UVW>$ or $<uvtw>$. For instance, $<100>$ in the cubic system signifies the directions [100], [010] and [001], the x-, y- and z-axes which are crystallographically equivalent.

A four-index system [$uvtw$] known as **Weber symbols** is sometimes used for directions in the hexagonal and trigonal systems. These symbols can be obtained from the Miller symbols [UVW] from the expressions:

$$u = \frac{2U - V}{3}; \quad v = \frac{2V - U}{3}; \quad t = -\frac{U + V}{3}; \quad w = W.$$

The reverse conversion (from four- to three-axis zone symbol) can be obtained from:

$$U = u - t; \quad V = v - t; \quad W = w.$$

In this book, both the three- and four-index zone symbols are given for trigonal and hexagonal zone directions.

Formulae for the angles between directions in the seven crystal systems can be found in Edington (1975) and may be calculated using many of the computer programs listed in Appendix F.

Note that, in the cubic system, a zone axis is parallel to the normal with the same indices. This is *not* generally true in any other crystal system.

A.4 Zones and the zone law

Any two lattice planes intersect in a line that can be described by indices [UVW]. A prism of planes also has a common direction and is known as a zone, the common direction being the **zone direction**. The indices of the zone direction common to two planes $h_1k_1l_1$ and $h_2k_2l_2$ can be found from:

$$U = k_1l_2 - k_2l_1; \quad V = l_1h_2 - l_2h_1; \quad W = h_1k_2 - h_2k_1.$$

For a plane (hkl) that lies at the intersection of two zones $[U_1V_1W_1]$ and $[U_2V_2W_2]$:

$$h = V_1W_2 - V_2W_1; \quad k = W_1U_2 - W_2U_1; \quad l = U_1V_2 - U_2V_1.$$

If you need to know if a plane (hkl) lies in a zone $[UVW]$ the condition is:

$$hU + kV + lW = 0.$$

This is called the **zone equation** or the **Weiss zone law.**

If the four-axis Weber and Miller–Bravais symbols are used the equation becomes:

$$hu + kv + it + lw = 0.$$

A.5 Software

Below are described three interactive software packages that are designed as aids to the understanding of crystallography.

Crystallography assumes no prior knowledge of basic crystallography and does not use the stereographic projection. The eight modules are: Introduction, Symmetry I, Symmetry II, Indexing I, Indexing II, Indexing III, Crystal Systems, Rotations. Available in PC and Macintosh formats from: The Earth Sciences Courseware Consortium, Department of Earth Sciences, University of Manchester, Manchester M13 9PL, U.K. (Tel: 44 (0)161 275 3820; Fax: 44 (0)161 275 3947; e-mail: ukescc@man.ac.uk; web-site: www.man.ac.uk/~ukescc.html).

Introduction to Crystallography also assumes no prior knowledge. The five sections are: Basic crystals, 2-D Crystallography, 3-D Crystallography, Crystal Structures, Indexing Directions, Planes. Available in PC format from: MATTER Coordinating Office, Department of Materials Science and Engineering, University of Liverpool, Liverpool L69 3BX, U.K. (Tel: 44 (0)151 794 5006; Fax: 44 (0)151 794 4675; e-mail: matter@liv.ac.uk; web-site: http:www.liv.ac.uk/~matter/home.html).

Ca.R.Ine Crystallography (version 3.1) is interactive program for teaching and research that covers real and reciprocal lattices, space groups, stereographic projections, planes and directions. Available in English and French in PC and Macintosh formats from: 17 rue du Moulin du Roy, F-60300 Senlis, France (e-mail: Cyrill.Boudias@UCT.fr *or* dmonceau@ENSCT.fr).

A.6 Further reading

Bloss, F.D. (1994) *Crystallography and Crystal Symmetry*. Mineralogical Society of America, Washington.

Borchardt-Ott, W. (1995) *Crystallography*, 2nd Edn. Springer Verlag, Berlin.

Glusker, J.P., Lewis, M. and Rossi, M. (1994) *Crystal Structure Analysis for Chemists and Biologists*. VCH publishers, New York.

Hammond, C. (2001) *The Basics of Crystallography and Diffraction*, 2nd Edn. Oxford University Press, Oxford.

Klein, K. and Hurlbut, Jr., C.S. (1993) *Manual of Mineralogy*. John Wiley, New York.

Steadman, R. (1982) *Crystallography*. Van Nostrand Reinhold, New York.

Appendix B

An introduction to space groups

B.1 Screw axes and glide planes

In the three-dimensional atomic structure of crystals there are two types of symmetry element that combine a symmetry operation (rotation or reflection) with a translation. A rotation axis with a translation parallel to the axis that is some fraction of the lattice repeat, t, is known as a **screw axis**, and a mirror reflection with a translation $t/2$ or $t/4$ parallel to the reflection plane is known as a **glide plane** or a **glide reflection**. The two-, three-, four-, and six-fold rotation axes can be combined with a translation (a one-fold rotation (not shown in *Figure B.1*) combined with a translation is equivalent to a translation only). *Figure B.1* and *Table B.1* show all the possible screw axes, along with their non-translational equivalents. The figure also shows the international graphical and written symbols for the axes, as used in the *International Tables for Crystallography* (Hahn, 1984). Screw axes are said to be **isogonal** (from the Greek meaning 'same angle') with the equivalent rotational axes. This means that four-fold screw axes, for example, rotate the motif through 90°, while translating it parallel to the axis. Screw axes are represented by the general symbol N_m, where N represents the rotation (two, three, four or six) and m represents the pitch in terms of the fraction of the translation t that is inherent in the operation. For example, the translation involved in the 4_1 screw axis is $\frac{1}{4}t$ (obtained by dividing the subscript, m, by the main axis symbol, N). There are three four-fold screw axes: 4_1, 4_2 and 4_3, with translations of $\frac{1}{4}t$, $\frac{1}{2}t$ and $\frac{3}{4}t$, respectively. By convention, when m/N is less than $\frac{1}{2}$, the screw is right handed (a right-handed screw is defined as one that advances away from the observer when rotated clockwise). Thus, 4_1 is a right-handed screw axis with translation $\frac{1}{4}t$ and 4_3 is a left-handed screw axis, with translation $-\frac{1}{4}t = \frac{3}{4}t$; the pair are said to be **enantiomorphous**. The pairs of screw axes 3_1 and 3_2, 6_1 and 6_5, 6_2 and 6_4 are also enantiomorphous. If m/N

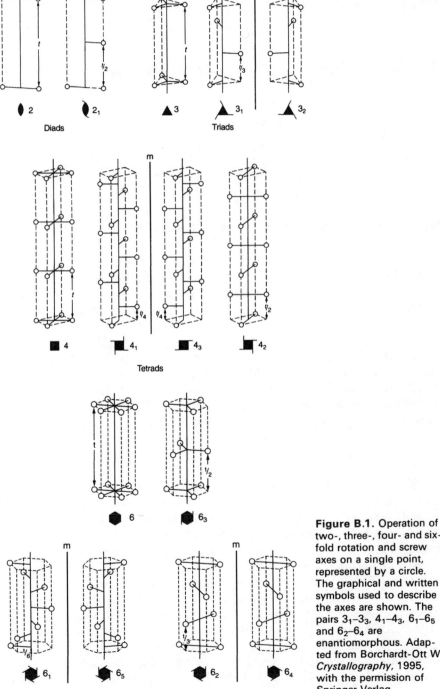

Figure B.1. Operation of two-, three-, four- and six-fold rotation and screw axes on a single point, represented by a circle. The graphical and written symbols used to describe the axes are shown. The pairs 3_1–3_3, 4_1–4_3, 6_1–6_5 and 6_2–6_4 are enantiomorphous. Adapted from Borchardt-Ott W, *Crystallography*, 1995, with the permission of Springer Verlag.

Table B.1. Symbols for rotation, rotoinversion and screw axes[1]

Symbol	Symmetry axis	Graphic symbol	Type of translation (if present)
1	One-fold rotation	None	None
$\bar{1}$	One-fold rotoinversion	○	None
2	Two-fold rotation	(parallel to paper)	None
2_1	Two-fold screw	(parallel to paper)	$\frac{1}{2}\mathbf{c}$ $\frac{1}{2}\mathbf{a}$ or $\frac{1}{2}\mathbf{b}$
3	Three-fold rotation	▲	None
3_1	Three-fold screw (right handed)		$\frac{1}{3}\mathbf{c}$
3_2	Three-fold screw (left handed)		$\frac{2}{3}\mathbf{c}$
$\bar{3}$	Three-fold rotoinversion	▲	None
4	Four-fold rotation	■	None
4_1	Four-fold screw (right handed)		$\frac{1}{4}\mathbf{c}$
4_2	Four-fold screw (neutral)		$\frac{2}{4}\mathbf{c} = \frac{1}{2}\mathbf{c}$
4_3	Four-fold screw (left handed)		$\frac{3}{4}\mathbf{c}$
$\bar{4}$	Four-fold rotoinversion		None
6	Six-fold rotation	●	None
6_1	Six-fold screw (right handed)		$\frac{1}{6}\mathbf{c}$
6_2	Six-fold screw (right handed)		$\frac{2}{6}\mathbf{c} = \frac{1}{3}\mathbf{c}$
6_3	Six-fold screw (neutral)		$\frac{3}{6}\mathbf{c} = \frac{1}{2}\mathbf{c}$
6_4	Six-fold screw (left handed)		$\frac{4}{6}\mathbf{c} = \frac{2}{3}\mathbf{c}$
6_5	Six-fold screw (left handed)		$\frac{5}{6}\mathbf{c}$
$\bar{6}$	Six-fold rotoinversion		None

[1]All shown normal to the page, unless otherwise indicated.

is equal to ½ i.e. 2_1, 4_2 and 6_3, the screw is considered to be neutral in direction.

If the translation produced by a glide plane is parallel to the x-axis, it is referred to as an *a*-glide (translation **a**/2) and is represented by the symbol *a*. Similarly, if the translation is parallel to the y- or z-axis, the glide is referred to as *b*- or *c*-glide, respectively (*Figure B.2* and *Table B.2*). If the translation is **a**/2 + **b**/2, **b**/2 + **c**/2 or **a**/2 + **c**/2, the operation is known as a **diagonal glide plane** and is represented by the symbol *n*. If the translation is **a**/4 + **b**/4, **b**/4 + **c**/4 or **a**/4 + **c**/4, the operation is known as a

Table B.2. Symbols for reflection planes and glide planes

Symbol	Symmetry plane	Normal to plane of projection	Parallel to plane of projection[1]	Nature of glide translation
m	Mirror	——————	⌐ ⌐ 120°	None
a, b		- - - - - - - -	⌐ ← ⌐	$a/2$ along the x axis or $b/2$ along the y axis
c	Axial glide plane	· · · · · · · · ·	None	$c/2$ along the z axis
n	Diagonal glide plane	·—·—·—·—·	⟋	$a/2 + b/2$; $a/2 + c/2$; $b/2 + c/2$; or $a/2 + b/2$; $+ c/2$ (tetragonal and cubic)
d	Diamond glide plane	·—·→·—·—· ·—·←·—·	1/8 ⟋ 3/8 ⟍	$a/4 + b/4$; $b/4 + c/4$; $a/4 + c/4$; or $a/4 + b/4$; $+ c/4$ (tetragonal and cubic)

[1]When planes are parallel to the paper, heights other than zero are indicated by writing the vertical coordinate next to the symbol (e.g. 1/4 or 3/8). The arrows indicate the direction of the glide.

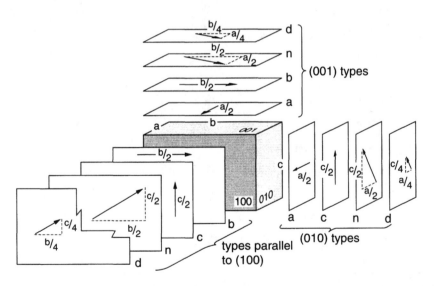

Figure B.2. Glide planes in an orthorhombic crystal whose unit cell is shown shaded. Reproduced from Bloss FD, *Crystallography and Crystal Chemistry*, 1994, with the permission of the Mineralogical Society of America.

diamond glide plane and is represented by the symbol d. *Table B.2* gives the standard written and graphical symbols used for glide and mirror planes.

B.2 Space group nomenclature

If the 14 Bravais lattices are combined with the symmetry inherent in the 32 point groups (non-translational symmetry), as well as the various glide planes and screw axes, we arrive at the 230 **space groups**. Each space group is isogonal with one of the 32 point groups. In other words, the point group is derived by replacing the translational symmetry elements in the space group with their non-translational equivalents. As an example, the space group illustrated in *Figure 5.5*, $P4_2/m\ 2_1/n\ 2/m$ belongs to point group $4/mmm$ (full symbol $4/m\ 2/m\ 2/m$), as indicated in the top right-hand corner of *Figure 5.5*. Another space group that is isogonal with point group $4/m\ 2/m\ 2/m$ is $I\ 4/m\ 2/c\ 2/m$. There are, in fact, 20 space groups that are isogonal with this point group.

Space groups are usually described, as above, by their **Hermann–Mauguin** symbols because they are easy to relate to the corresponding point group, although the *International Tables for Crystallography* (Hahn, 1984) also gives the **Schoenflies** symbol for each entry underneath the Hermann–Mauguin symbol (at the top left in *Figure 5.5*; D_{4h}^{14} in this case). In the Hermann–Mauguin symbol, the symmetry elements are preceded by the lattice type: P and I, respectively, in the two examples above.

One point about the notation for point groups and space groups that may be worth clarifying is the use by crystallographers of what are known as 'abbreviated' symbols. The space group $P4_2/m\ 2_1/n\ 2/m$ and its isogonal point group $4/m\ 2/m\ 2/m$ are more usually written in the abbreviated forms: $P4_2/mnm$ and $4/mmm$, respectively. The reasoning behind the use of the abbreviated symbols is that they show the 'essential' symmetry of the point group or space group; for instance, the intersection of mirror planes every $45°$ perpendicular to the tetrad (as signified by the second and third m in the point group $4/mmm$) dictates that there are also diads at the intersections of these mirror planes. Note that the presence of centres of symmetry is *never* indicated in point or space group symbols, even in the 'full' versions.

B.3 Space group drawings

Each space group is a scaffold-like array of symmetry elements. The *International Tables for Crystallography* (Hahn, 1984) projects the appearance of each space group onto the x–y plane; that is onto (001). The y-axis is oriented sideways, with the positive direction to the right, the x-axis extends down the page and the z-axis is vertical, towards the reader. The conventional choice of origin is given and is always a centre of symmetry, if one is present. In *Figure 5.5* the statement reads: "Origin at centre (mmm)".

There are two drawings of each space group. The drawing on the right shows one unit cell with the symmetry elements indicated using the symbols given in *Tables B.1* and *B.2*. The second drawing shows the operation of the symmetry elements on an asymmetric motif or group of atoms represented by a small circle. The circle is placed at small fractions x, y, z of the unit cell edges away from the origin, with the height $(+)z$

indicated by a + sign. When a symmetry operation such as a reflection plane or centre of symmetry changes the hand of the motif, the circle has a comma inside it. When two motif positions project on top of one another, the circle representing them is split in two by a vertical line, the circle's left side representing one atom and its right side representing the other, as in *Figure 5.5*.

The set of positions generated by the symmetry on the 'starting' position is known as the set of **general equivalent positions** because the positions do not lie on any of the symmetry elements. The coordinates of these positions are listed below the drawings, together with their total number (16 in *Figure 5.5*), their Wyckoff letter, *k* in this case, and the symmetry of the position, 1 (no symmetry).

Below the general equivalent positions are listed the **special equivalent positions**. These are positions whose number is less than the general equivalent positions because they lie on non-translational symmetry elements, i.e. rotation and (roto)inversion axes, mirror planes and centres of symmetry, and the motif is not repeated by these symmetry elements. In *Figure 5.5*, the number varies from 8 for positions *j* and *i* that lie on a single mirror plane, *m*, to 2 for positions *b* and *a* at the intersections of three mirror planes, *mmm.* (note that this is also where the centres of symmetry are).

The right-hand column in *Figure 5.5* lists the conditions limiting possible reflections in diffraction patterns, and the first group of entries (headed General) apply to any crystal with this space group. The special conditions listed below the general conditions only apply if *all* the atoms are in the special positions in the unit cell whose coordinates are listed in the central column. This topic is dealt with in Sections 5.2.1 and 5.2.3 of the main text.

B.4 Further reading

Bloss, F.D. (1994) *Crystallography and Crystal Symmetry*. Mineralogical Society of America, Washington.

Borchardt-Ott, W. (1995) *Crystallography*, 2nd Edn. Springer Verlag, Berlin.

Hahn, T. (1984) *International Tables for Crystallography. Volume A, space group symmetry*. Kluwer Academic Publishers, Dordrecht, The Netherlands.

Hammond C. (2001) *The Basics of Crystallography and Diffraction*, 2nd Edn. Oxford University Press, Oxford.

Klein, K. and Hurlbut, Jr., C.S. (1993) *Manual of Mineralogy*. John Wiley, New York.

Appendix C

Some useful crystallographic relationships

C.1 Interplanar spacings, d_{hkl}, and the relationship between direct, a, b, c, α, β, γ, and reciprocal, a^*, b^*, c^*, α^*, β^*, γ^*, lattice parameters

Orthorhombic:

$a^* = 1/a$; $b^* = 1/b$; $c^* = 1/c$; $\alpha^* = \beta^* = \gamma^* = 90°$

$$\frac{1}{d_{hkl}^2} = \frac{h^2}{a^2} + \frac{k^2}{b^2} + \frac{l^2}{c^2}$$

Tetragonal: $a = b$; cubic: $a = b = c$
Hexagonal and trigonal (hexagonal axes):

$a^* = b^* = 2/(a\sqrt{3})$; $c^* = 1/c$; $\alpha^* = \beta^* = 90°$; $\gamma^* = 60°$

$$\frac{1}{d_{hkl}^2} = \frac{4}{3}\left(\frac{h^2 + hk + k^2}{a^2}\right) + \frac{l^2}{c^2}$$

Trigonal (rhombohedral axes):

$a = b = c$; $\alpha = \beta = \gamma \neq 90°$

$a^* = b^* = c^* = (a^2 \sin \alpha)/V$; $\cos \alpha^* = (\cos^2 \alpha - \cos \alpha)/\sin^2\alpha$

where V = volume of the unit cell = $a^3\sqrt{1 - 3\cos^2 \alpha + 2\cos^3 \alpha}$

$$\frac{1}{d_{hkl}^2} = \frac{(h^2 + k^2 + l^2)\sin^2\alpha + 2(hk + kl + hl)(\cos^2\alpha - \cos\alpha)}{a^2(1 - 3\cos^2\alpha + 2\cos^3\alpha)}$$

The rhombohedral lattice can be reduced to the (triple) hexagonal description a', b':

$$a' = 2a\sin(\alpha/2), \quad c' = a\sqrt{3}\sqrt{(1 + 2\cos\alpha)}$$

Monoclinic:

$$a^* = 1/(a\sin\beta); \quad b^* = 1/b; \quad c^* = 1/(c\sin\beta); \quad \alpha^* = \gamma^* = 90°; \quad \beta^* = 180 - \beta$$

$$\frac{1}{d_{hkl}^2} = \frac{1}{\sin^2\beta}\left(\frac{h^2}{a^2} + \frac{k^2\sin^2\beta}{b^2} + \frac{l^2}{c^2} - \frac{2hl\cos\beta}{ac}\right)$$

Triclinic:

$$a^* = (bc\sin\alpha)/V; \quad b^* = (ac\sin\beta)/V; \quad c^* = (ab\sin\gamma)/V$$

$$\cos\alpha^* = (\cos\beta\cos\gamma - \cos\alpha)/(\sin\beta\sin\gamma); \quad \cos\beta^* = (\cos\gamma\cos\alpha - \cos\beta)/(\sin\alpha\sin\gamma); \quad \cos\gamma^* = (\cos\alpha\cos\beta - \cos\gamma)/(\sin\alpha\sin\beta)$$

$$\frac{1}{d^2} = \frac{1}{V}(S_{11}h^2 + S_{22}k^2 + S_{33}l^2 + 2S_{12}hk + 2S_{23}kl + 2S_{13}hl)$$

where V = volume of the unit cell

$$= abc\sqrt{1 - \cos^2\alpha - \cos^2\beta - \cos^2\gamma + 2\cos\alpha\cos\beta\cos\gamma}$$

$$S_{11} = b^2c^2\sin^2\alpha \qquad S_{12} = abc^2(\cos\alpha\cos\beta - \cos\gamma)$$

$$S_{22} = a^2c^2\sin^2\beta \qquad S_{23} = a^2bc(\cos\beta\cos\gamma - \cos\alpha)$$

$$S_{33} = a^2b^2\sin^2\gamma \qquad S_{13} = ab^2c(\cos\gamma\cos\alpha - \cos\beta)$$

C.2 Interplanar angles, ρ, between planes $(h_1k_1l_1)$ and $(h_2k_2l_2)$

Orthorhombic:

$$\cos\rho = \frac{(h_1h_2/a^2) + (k_1k_2/b^2) + (l_1l_2/c^2)}{\sqrt{\{[(h_1^2/a^2) + (k_1^2/b^2) + (l_1^2/c^2)][(h_2^2/a^2) + (k_2^2/b^2) + (l_2^2/c^2)]\}}}$$

Tetragonal: $a = b$; *cubic*: $a = b = c$

Hexagonal and trigonal (hexagonal axes):

$$\cos\rho = \frac{h_1h_2 + k_1k_2 + \frac{1}{2}(h_1k_2 + h_2k_1) + (3a^2/4c^2)l_1l_2}{\sqrt{\{[h_1^2 + k_1^2 + h_1k_1 + (3a^2/4c^2)l_1^2][h_2^2 + k_2^2 + h_2k_2 + (3a^2/4c^2)l_2^2)]\}}}$$

Trigonal (rhombohedral axes):

$$\cos \rho = (a^4d_1d_2/V^2)[\sin^2 \alpha(h_1h_2 + k_1k_2 + l_1l_2)$$
$$+ (\cos^2 \alpha - \cos \alpha)(k_1l_2 + k_2l_1 + l_1h_2 + l_2h_1 + h_1k_2 + h_2k_1)]$$

Monoclinic:

$$\cos \rho = \frac{d_1d_2}{\sin^2 \beta}\left[\frac{h_1h_2}{a^2} + \frac{k_1k_2 \sin^2 \beta}{b^2} + \frac{l_1l_2}{c^2} - \frac{(l_1h_h + l_2h_1) \cos \beta}{ac}\right]$$

Triclinic:

$$\cos \rho = \frac{d_1d_2}{V^2} [S_{11}h_1h_2 + S_{22}k_1k_2 + S_{33}l_1l_2 + S_{23}(k_1l_2 + k_2l_1)$$
$$S_{13}(l_1h_2 + l_2h_1) + S_{12}(h_1k_2 + h_2k_1)]$$

C.3 Angles between planes (or zones) in the cubic system

Note that, in the cubic system, a zone axis is parallel to the plane normal with the same indices. This is *not* generally true in any other crystal system. The indices below represent forms or sets of symmetrically equivalent zones. Therefore, their indices may be permuted, together with their negative counterparts.

100	100	0.00	90.00			
	110	45.00	90.00			
	111	54.74				
	210	26.56	63.43	90.00		
	211	35.26	65.90			
	221	48.19	70.53			
110	110	0.00	60.00	90.00		
	111	35.26	90.00			
	210	18.43	50.77	71.56		
	211	30.00	54.74	73.22	90.00	
	221	19.47	45.00	76.37	90.00	
111	111	0.00	70.53			
	210	39.23	75.04			
	211	19.47	61.87	90.00		
	221	15.79	54.74	78.90		
210	210	0.00	36.87	53.13	66.42	78.46	90.00
	211	24.09	43.09	56.79	79.48	90.00
	221	26.56	41.81	53.40	63.43	72.65	90.00
211	211	0.00	33.56	48.19	60.00	70.53	80.40
	221	17.72	35.26	47.12	65.90	74.21	82.18

Appendix D

Relativistic electron wavelengths

Accelerating voltage, V(kV)	Wavelength[1] λ (pm)
20	8.588
30	6.978
40	6.014
50	5.355
60	4.865
70	4.485
80	4.177
90	3.919
100	3.702
200	2.508
300	1.968
400	1.644
500	1.421
600	1.256
700	1.129
800	1.027
900	0.942
1000	0.872

[1] 1 pm = 10^{-12} m = 10^{-3} nm = 10^{-2} Å.

Wavelengths were calculated using the equation:

$$\lambda = [h\sqrt{2emV}] \times (1+9.78 \times 10^{-7}\ V)^{-1/2}$$

where h (Planck's constant) = 6.626×10^{-34} Js; e (charge on the electron) = 1.602×10^{-19} C; m (relativistically corrected mass of electron) = $m_0 (1 - \frac{2eV}{m_0 c^2})^{-1/2}$, where m_0 (rest mass of electron) = 9.110×10^{-31} kg and c (the velocity of light) = 2.998×10^8 m/s. V is in volts and λ is in metres.

Appendix E

Mathematical definition of the reciprocal lattice

For a simple introduction to vectors and complex numbers and their use in crystallography, see Hammond (2001).

$\mathbf{a}$, $\mathbf{b}$ and $\mathbf{c}$ are the unit-cell dimensions of the direct lattice and $\mathbf{a^*}$, $\mathbf{b^*}$ and $\mathbf{c^*}$ are the dimensions of the reciprocal lattice. These vectors are related by the relations:

$$\mathbf{a^*.b} = \mathbf{a^*.c} = \mathbf{b^*.c} = \mathbf{b^*.a} = \mathbf{c^*.a} = \mathbf{c^*.b} = 0.$$

Also:

$$\mathbf{a^*.a} = \mathbf{b^*.b} = \mathbf{c^*.c} = 1$$

The volume of the direct unit cell V_c is given by $\mathbf{a.(b \times c)}$ and the volume of the reciprocal unit cell is given by $\mathbf{a^*.(b^* \times c^*)}$:

$$\mathbf{a^*} = \frac{\mathbf{b \times c}}{V}, \ \mathbf{b^*} = \frac{\mathbf{c \times a}}{V}, \ \mathbf{c^*} = \frac{\mathbf{a \times b}}{V};$$

$$\mathbf{a} = \frac{\mathbf{b^* \times c^*}}{V}, \ \mathbf{b} = \frac{\mathbf{c^* \times a^*}}{V}, \ \mathbf{c} = \frac{\mathbf{a^* \times b^*}}{V}.$$

Appendix F

Computer programs concerning electron diffraction

There are a number of computer programs available that deal with electron diffraction. Some are designed as teaching aids, but most are designed as aids to the interpretation of diffraction patterns. Below is a list of those known to the author, but it is not guaranteed to be exhaustive! I have only listed the functions relevant to electron diffraction; many of the programs have a number of other functions. It is also worth noting that this field changes very fast! All programs run on PCs unless otherwise stated.

DIFFRACTION (P.J. Goodhew, A. Fretwell, B. Tanovic, I. Jones, A. Green, D. Brook) Free, interactive, on-line educational software from the MATTER team covering an introductory undergraduate course on diffraction (including electron diffraction). Web-site: www.matter.org.uk/diffraction/

ELECTRON DIFFRACTION version 6.8 (J.P. Morniroli.) Simulates spot, ring, Kikuchi and CBED patterns (including large-angle patterns and HOLZ lines). Automatic interpretation of spot patterns from any crystal system other than triclinic. The space group, lattice parameters, atomic species and atomic positions can be input. Crystallographic calculations. e-mail: jean-paul.morniroli@univ-lille1.fr; web-site: www.univ-lille1.fr/lmpgm

IDEALMICROSCOPE (J.M. Zuo, S.H.R. Zhu) for Macintosh. Draws and indexes electron diffraction patterns, including CBED patterns, from crystals of any symmetry. HOLZ and Kikuchi lines can also be plotted. Input data are cell parameters, atomic species and atomic positions. Interactive: changes in patterns can be watched as tilts and other microscope parameters are manipulated. Web-site: www.microscopy-online.com/Vendors/EMLab

JEMS (P. Stadelmann). A Java version of older **EMS** software suitable for PCs and Macintoshs. Simulates kinematic and dynamical spot diffraction patterns, Kikuchi lines and maps, and CBED (with HOLZ lines). Will index experimental patterns. Input data are cell parameters, space group, atomic species and Wyckoff positions. A version is available free over the Internet. e-mail: pierre.stadelmann@epfl.ch; web-site: cimewww.epfl.ch

WINDOWS CM k-SPACE CONTROL. An on-line program for indexing diffraction patterns and orienting crystals in Philips/FEI TEMs. Input data are cell parameters and lattice type. Also shows zone axes and Kikuchi maps, and can instruct TEM stage to tilt to the desired zone axis automatically. Web-site: www.feic.com/support/cmks.html

MEETINGS AND COURSES. The most comprehensive web-site for information about courses and meetings on electron microscopy all over the world is run by Nestor Zaluzec at Argonne National Laboratory, USA (www.amc.anl.gov). It also has links to microscopy societies world-wide. Another useful source of information on societies and software is the Electron Micoscopy Yellow Pages: cimewww.epfl.ch.emyp/comp.html

Appendix G

Answers to exercises

Chapter 2

2.1 The scattering angle α in the equation is equal to 2θ, θ being the Bragg angle for the reflection. θ is obtained from the equation $\theta \sim \lambda/2d$. $d_{222} = a/\sqrt{12}$ (see Appendix C) $= 0.117$ nm, so $\theta = 0.016$ radians. This gives a displacement of 98 nm. The Bragg angle at 300 kV is 0.0085 radians and the displacement in the object plane is 5 nm.

2.2 *Figure 2.22a* is [001] and *Figure 2.22b* is [011]. The lattice parameter is 0.404 nm.

2.3 The camera constant is 2.9 nm mm.

2.4 The values are $a = 0.48$ nm and $b = 1.02$ nm.

2.5 Because the lattice is C-face centred, $h + k = 2n$, where n is an integer. For the [010] pattern therefore, as $k = 0$, $h = 2n$ (*Figure G.1*). The angle β^* is 78°, so $\beta = 180 - \beta^* = 102°$. $R_{200} = 5.17$ mm, so $d_{200} = \lambda L/R = 0.46$ nm. As $d_{200} = a \sin \beta/2$, $a = 0.94$ nm. Similarly $d_{001} = c\sin\beta$ and $c = 0.53$ nm.

2.7 Reflections are absent for a face-centred lattice if h, k and l are not all odd or all even. This leads to the reciprocal unit cell shown in *Figure G.2a*. For the C-face centred lattice, reflections are absent if $h + k \neq 2n$. This leads to the reciprocal unit cell shown in *Figure G.2b*.

Chapter 3

3.1 The matrix and precipitate patterns are both from $<110>$ zones and the 200_{matrix} and 400_{ppt} spots coincide. Therefore, the orientation relationship is: $(100)_{\alpha Fe}//(100)_{ppt}$; $[0\bar{1}1]_{\alpha\text{-Fe}}//[0\bar{1}1]_{ppt}$. This is the 'cube–cube' relationship and could be written in a large number of other

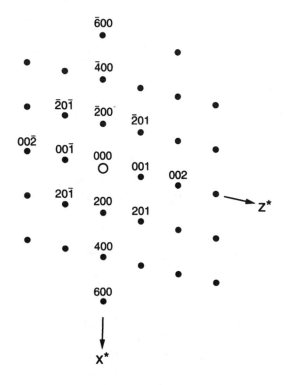

Figure G.1.

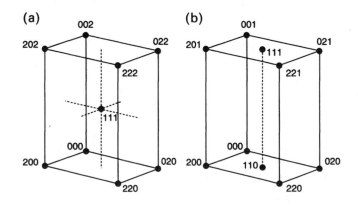

Figure G.2.

ways because any two {hkl} planes and <UVW> directions in the cubic system are parallel. An even simpler way of describing the relationship is: $(100)_{\alpha\text{-Fe}}//(100)_{\text{ppt}}$; $[010]_{\alpha\text{-Fe}}//[010]_{\text{ppt}}$. As the 200_{matrix} and 400_{ppt} spots coincide, the d-values for the corresponding planes must be equal and $a_{\text{ppt}} = 2a_{\text{matrix}}$. So the lattice parameter of the precipitate is 0.5732 nm.

3.2 The matrix zone axis is <011> and that of the precipitate is <001>. The matrix $1\bar{1}1$ and precipitate 110 spots coincide (*Figure G.3*).

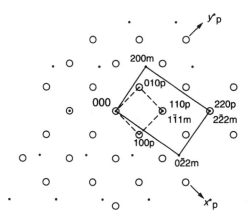

Figure G.3.

Therefore, the orientation relationship is: $(1\bar{1}1)_{\text{matrix}}//(110)_{\text{ppt}}$; $[011]_{\text{matrix}}//[001]_{\text{ppt}}$. This is known as the Nishiyama–Wassermann relationship.

Chapter 4

4.1 As $g^2 = (1/d)^2 = (h^2 + k^2 + l^2)/a^2$, for the (222) planes $g = 10.97$ nm^{-1}. (b) The beam is in the symmetry condition with respect to the ± 222 Kikuchi lines, so $s = g^2 \lambda/2 = 0.223$ nm^{-1}. (c) From the diffraction pattern $R = 10.5$ mm and $\Delta R = 2.5$ mm. $s = \Delta R/R \times g^2 \lambda = 0.10$ nm^{-1}. s is positive, as noted earlier. (d) $\Delta R = 2$ mm and $s = 0.08$ nm^{-1}.

4.2 The pattern shows four-fold symmetry; therefore the zone axis is $<001>$. Because of the extra systematic absences, the spot diffraction pattern will resemble the [001] pattern for diamond cubic (see *Figure 2.16*) and the shortest **g**-vectors, and hence the narrowest Kikuchi bands, will be those for the {220} planes. The two narrower Kikuchi bands at approximately 45° to the edges of the page are therefore ± 220, and $\pm 2\bar{2}0$ and the wider bands at 45° to the narrow ones are ± 400 and ± 040. The width of the 220 Kikuchi band is 10 mm, which, using the camera constant, gives a value of $a = 0.81$ nm. You would need to tilt along the ± 400 or ± 040 Kikuchi band to reach the nearest $<110>$ zone axis. From [001], the zone would be [011] or [101] as the planes (400) and (040) are parallel to these zones, respectively (Weiss zone law). The angle of tilt would be 45°.

4.3 The d-value for the (200) planes is 0.202 nm, which gives $\theta_{200} = 0.525°$. The width of the 200 Kikuchi band, R_{200}, is 8.2 mm. The conversion factor from distance to angle is $2\theta/R = 0.128$ (see Section 4.4). The values of X and Y in *Figure G.4* are 8.5 and 7 mm, respectively; these values translate to angles of 1.09° about the normal to the (200) planes and 0.90° about the normal to the (02$\bar{2}$) planes. Using the formula given in Section 4.4, the resultant tilt is 1.4°.

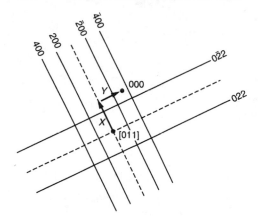

Figure G.4.

Chapter 5

5.1 (a) An atom related to an atom in a general position x_n, y_n, z_n by a screw diad parallel to the y-axis would have the coordinates $-x_n$, $1/2+y_n$, $-z_n$.

(b) The structure factor is:

$$F_{hkl} = \sum_{n=1}^{N/2} f_n \left[\exp 2\pi i (hx_n + ky_n + lz_n) + \exp 2\pi i \left(-hx_n + ky_n + \frac{k}{2} - lz_n\right) \right].$$

If h and l are zero:

$$F_{0k0} = \sum_{n=1}^{N/2} f_n \left[exp2\pi i ky_n + exp2\pi i \left(ky_n + \frac{k}{2}\right) \right].$$

If k is odd:

$$F_{0k0} = 0.$$

So $0k0$ reflections are absent if k is odd.

(c) Physically, by only considering the $0k0$ reflections we are projecting the atoms onto the y-axis, which is effectively halved; so the repeat in reciprocal space is doubled.

(d) Absences in the $0k0$ reflections would also occur if there were an n- or a b-glide plane parallel to (001) or (100). However, in both cases there would be other absences in addition to those for $0k0$ reflections. For instance, an n-glide plane parallel to (001) would result in absences for $hk0$ reflections when $h + k \neq 2n$.

5.2 Absences in the [010] pattern for l odd would also occur if the lattice were B-face centred. The two possibilities could be distinguished by tilting the specimen to any other zone, when the general absences in the *hkl* reflections for the B-face-centred lattice ($h + l \neq 2n$) would be apparent.

Chapter 6

6.1 The radius of the FOLZ is 25.5 mm. The formula for the zone-axis repeat, assuming $N = 1$ (P lattice), gives a value of $c = 1.5$ nm, which is the repeat for the 6H polytype. The BF (full) and WP (full) symmetries are both *6mm* (the WP (proj) and BF (proj) symmetries are difficult to see). From *Table 6.1*, the diffraction group is either *6mm*1_R or *6mm*. From *Figure 6.8*, the possible point groups are *6/mmm* or *6mm*.

6.2 The radius of the FOLZ is 19.75 mm. The formula for the zone-axis repeat using $N = 1$ gives a value of $b = 0.9$ nm.

6.3 The [100], [010] and [001] diffraction patterns are all rectangles or centred rectangles, so the crystal system is orthorhombic. This is confirmed by *Figure 6.13(a)*, in which the FOLZ clearly stacks vertically above the ZOLZ. In the [001] pattern the distribution of the reflections in the ZOLZ and the FOLZ is the same. Thus, there is no glide plane parallel to (001). In the [100] pattern the reflections for which $k + l \neq 2n$ are absent in the ZOLZ, but present in the FOLZ. This behaviour indicates the presence of an *n*-glide plane parallel to (100). In the [010] pattern the $h \neq 2n$ reflections are absent in the ZOLZ, but present in the FOLZ. This indicates that there is an *a*-glide plane parallel to (010). The diffraction symbol is therefore *Pna*-, where the - indicates an undetermined symmetry element: either a mirror plane or a diad. The point group is either *mm2* or *mmm* (see *Table A.1*). These point groups can be distinguished by the CBED pattern in *Figure 6.13d*. The WP (full) symmetry is *2mm*. From *Table 6.3*, point group *mm2* shows diffraction group symmetry *m*1_R along [100], while point group *mmm* shows diffraction group symmetry *2mm*1_R. From *Table 6.2*, *m*1_R corresponds to WP (full) symmetry *m*, while *2mm*1_R corresponds to WP (full) symmetry *2mm*. Thus, the point group is *mmm* and the space group is *Pnam* = $P2_1/n\ 2_1/a\ 2_1/m$. This is the same space group as olivine (see *Figure 6.5*), but with the crystal axes specified in a different order. The *International Tables for Crystallography* (Hahn, 1984) gives tables of the space-group symbols for all the possible settings of the unit cell.

6.4 The WP (full) symmetry is *4mm*. From *Table 6.2*, the possible diffraction groups are *4mm* and *4mm*1_R. From *Figure 6.8*, these diffraction groups are consistent with the tetragonal point groups *4/mmm* and *4mm*, and the cubic point group *m3m*. Note that other ZAP symmetries at this zone axis would not help to narrow the possibilities in this case (see *Tables 6.1* and *6.2*). Tilting to another zone axis, e.g. [111], would distinguish between them, however. Measurement of the FOLZ ring in *Figure 6.2b* would give the repeat parallel to the beam and indicate whether the system was cubic or tetragonal (if it is cubic the value must be the same as measured for a and b from the conventional diffraction pattern; if it is tetragonal it would most likely be different). Note that the equality $c = a = b$ is a necessary, *but not sufficient,* condition for a crystal to belong to the cubic system. The symmetry must conform as well.

6.5   The possible diffraction groups (DG), as deduced from *Tables 6.1* and *6.2*, are shown in the fourth column of the table below and the hexagonal point groups with which they are consistent are also shown*. The only point group that is consistent with all the ZAPs is $6/mmm$.

Zone axis	WP (proj)	WP (full)	DG	Possible point group			
				$6/m$	$6mm$	$6m2$	$6/mmm$
[001]	$6mm$	$6mm$	$6mm$ or	*			
			$6mm1_R$				*
[1$\bar{1}$0]	$2mm$	$2mm$	$2mm$ or			*	
			$2mm1_R$				*
[1$\bar{1}$1]	$2mm$	m	2_Rmm_R	*			*

6.6   The BF (proj) symmetry (that of the broad detail) is $6mm$ and the BF (full) symmetry (that of the fine HOLZ lines) is $3m$. Using the other two symmetries given, the diffraction group is 6_Rmm_R and the two possible point groups are $m\bar{3}m$ and $\bar{3}m$.

Chapter 7

7.1   Screw axes always result in the systematic absences being confined to a single row in the diffraction pattern and there is therefore always a route for double diffraction to occur. For instance, a screw diad parallel to the y-axis produces absences in $0k0$ reflections when $k \neq 2n$. In a [100] diffraction pattern one route for double diffraction into the 010 reflection is by diffraction by the (011) planes and then re-diffraction by the (00$\bar{1}$) planes (or *vice versa*) because $011 + 00\bar{1} = 010$.

7.2   The streaking is along z*, so the faults are on the (001) planes. The streaking affects the reflections for which $k \neq 3n$, so the fault vector is $b/3$. As the streaking is continuous, the probability of a fault occurring is about 0.5 or 50%.

Appendix H

References

Azough, F., Champness, P.E. and Freer, R. (1995) Determination of the space group of ceramic $BaO.Pr_2O_3.4TiO_2$ by electron diffraction. *J. Appl. Cryst.* **28**: 577–581.

Bloss, F.D. (1994) *Crystallography and Crystal Symmetry*. Mineralogical Society of America, Washington.

Borchardt-Ott, W. (1995) *Crystallography*, 2nd Edn. Springer Verlag, Berlin.

Buxton, B.F., Eades, J.A., Steeds, J.W. and Rackham, G.M. (1976) The symmetry of electron diffraction zone axis patterns. *Phil. Trans Roy. Soc.* **281A**: 171–193.

Champness, P.E. (1987) Convergent beam electron diffraction. *Mineral. Mag.* **51**: 33–48.

Champness, P.E. (1995) Analytical electron microscopy. In: *Microprobe Techniques in the Earth Sciences* (eds P.J. Potts, J.W.F. Bowles, S.J.B. Reed and M.R. Cave). Chapman and Hall, London, pp. 91–139.

Chescoe, D.C. and Goodhew, P.J. (1990) *The Operation of the Transmission and Scanning Electron Microscopes*, RMS Microscopy Handbook 20. Oxford University Press, Oxford.

Cliff, G. and Kenway, P.B. (1982) The effects of spherical aberration in probe-forming lenses on probe size, image resolution and X-ray spatial resolution in scanning transmission electron microscopy. In: *Microbeam Analysis–1982* (ed. K.F.J. Heinrich). San Francisco Press Inc, San Francisco, pp. 107–110.

Eades, J.A. (1984) Zone-axis diffraction patterns by the "Tanaka" method. *J. Electron Microsc. Technique* **1**: 279–284.

Eades, J.A. (1988a) Symmetry determination by convergent-beam diffraction. *Proc. 9th European Congress Electron Microscopy*, York, 1988. IOP Publishing, Bristol, pp. 3–12.

Eades, J.A. (1988b) Glide planes and screw axes in convergent-beam diffraction: the standard procedure. *Microbeam Anal.* San Francisco Press Inc, San Francisco, pp. 75–80.

Eades, J.A. (1992) Coherent-beam diffraction. In: *Electron Diffraction Techniques*, Vol. 1 (ed. J.M. Cowley). International Union of

Crystallography Monograph 3, Oxford University Press, Oxford, pp. 313–359.

Edington, J.W. (1975) *Electron Diffraction in the Electron Microscope.* Macmillan, London.

Gard, A. (1971) *The Electron-optical Investigation of Clays.* Mineralogical Society Monograph 3, Mineralogical Society London.

Gjønnes, J. and Moodie, A.F. (1965) Extinction conditions in the dynamic theory of electron diffraction. *Acta Crystallog.* **19**: 65–67.

Glusker, J.P., Lewis, M. and Rossi, M. (1994) *Crystal Structure Analysis for Chemists and Biologists.* VCH publishers, New York.

Goodhew, P.J. (1984) *Specimen Preparation for Transmission Electron Microscopy.* Oxford University Press, Oxford.

Guinier, A. (1958) Heterogeneities in solid solutions. *Sol. State Phys.* **9**: 293–301.

Guinier, A. (1963) *X-ray Diffraction in Crystals, Imperfect Crystals and Amorphous Bodies.* W.H. Freeman and Co., San Francisco.

Hahn, T. (1984) *International Tables for Crystallography. Volume A, Space Group Symmetry.* Kluwer Academic Publishers, Dordrecht, The Netherlands.

Haider, M., Rose, H., Uhlemann, S., Schwan, E., Krabius, B. and Urban, K. (1998) A spherical-aberration corrected 200kV transmission electron microscope. *Ultramicroscopy* **75**: 53–60.

Hammond, C. (1993) Stereographic techniques, Module 9.17. In: *Procedures for Transmission Electron Microscopy* (eds A. Robards and A.J. Wilson). John Wiley and Sons, Chichester.

Hammond, C. (2001) *The Basics of Crystallography and Diffraction,* 2nd Edn. Oxford University Press, Oxford.

Hasan, F., Jahanafrooz, A., Lorimer, G.W. and Ridley, N. (1982) The morphology, crystallography and chemistry of phases of an as-cast nickel-aluminium bronze. *Metall. Trans.* **13**A: 1337–1345.

Henry, N.F.M. and Lonsdale, K. (1952) *International Tables for X-ray Crystallography. Volume I, Symmetry Groups.* Kynoch Press, Birmingham.

Hirsch, P.B., Howie, A., Nicholson, R.B. and Pashley, D.W. (1965) *Electron Microscopy of Thin Crystals.* Butterworths, London.

Keyse, R.J., Garratt-Reed, A.J., Goodhew, P.J. and Lorimer GW. (1998) *Introduction to STEM.* BIOS Scientific Publishers, Oxford.

Krivenek, O.L., Delby, N. and Lupini, A.R. (1999) Towards sub-Å electron beams. *Ultramicroscopy* **78**: 1–11.

Levine, E., Bell, W.L. and Thomas, G. (1966) Further applications of Kikuchi patterns: Kikuchi maps. *J. Appl. Phys.* **37**: 2141–2148.

Lewis, M.H. and Billingham, J. (1972) *JEOL News* **10**e(1): 8.

Loretto, M.H. (1994) *Electron Beam Analysis of Materials.* Chapman and Hall, London.

Lorimer, G.W. and Champness, P.E. (1974) Mineralogical applications of HVEM. In: *High Voltage Electron Microscopy* (ed. P.R. Swann). Academic Press, London, pp. 301–311.

Lyman, C.E. and Carr, M.J. (1993) Identification of unknowns. In: *Electron Diffraction Techniques*, Vol. 2 (ed. J.M. Cowley). International Union of Crystallography Monograph 4, Oxford University Press, Oxford, pp. 373–417.

MacGillavry, C.H. and Rieck, G.D. (1983) *International Tables for Crystallography. Volume D, Physical and Chemical Tables*. Kluwer Academic Publishers, Dordrecht, The Netherlands.

Misell, D.L. and Brown, E.B. (1987) *Electron Diffraction: an Introduction for Biologists*. Elsevier, Amsterdam.

Morniroli, J.P. and Steeds, J.W. (1992) Microdiffraction as a tool for crystal structure determination. *Ultramicroscopy* **45**: 219–239.

NIST Crystal Data. Obtainable on CD ROM from NIST Crystal Data Center, NIST, Gaithersburg, MD 20899, USA.

O'Keefe, M. and Hyde, B.G. (1996) *Crystal Structures. I. Patterns and Symmetry*. Mineralogical Society of America, Washington.

Owen, D.C, and McConnell, J.D.C. (1971) Spinodal decomposition in an alkali feldspar. *Nature Phys. Sci.* **230**: 118–120.

Pashley, D.W. and Presland, A.E.B. (1959) The observation of antiphase boundaries during the transition from CuAuI to CuAuII. *J. Inst. Metals* **87**: 419–428.

Pashley, D.W. and Stowell, M.J. (1963) Electron microscopy and diffraction of twinned structures in evaporated films of gold. *Phil. Mag.* **8**: 1605–1632.

Porter, D.A. and Easterling, K.E. (1992) *Phase Transformations in Metals and Alloys*, 2nd Edn. Chapman and Hall, London.

Putnis, A. (1992) *Introduction to Mineral Sciences*. Cambridge University Press, Cambridge.

Rymer, T.B. (1970) *Electron Diffraction*. Methuen, London.

Sauvage, M. and Parté, E. (1972) Vacancy short-range order in substoichiometric transition metal carbides and nitrides with the NaCl structure. II. Numerical calculation of vacancy arrangement. *Acta Cryst.* **A28**: 607–616.

Smith, P.P.K. (1978) Note on the space group of spinel minerals. *Phil. Mag.* **38**: 99–102.

Spence, J.C.H. (1992) Accurate structure factor amplitude and phase determination. In: *Electron Diffraction Techniques*, Vol. 1 (ed. J.M. Cowley). International Union of Crystallography Monograph 3, Oxford University Press, Oxford, pp. 360–438.

Spence, J.C.H. and Zuo, J.M. (1992) *Electron Microdiffraction*. Plenum Press, New York.

Steadman, R. (1982) *Crystallography*. Van Nostrand Reinhold, New York.

Steeds, J.W. (1979) Convergent beam electron diffraction. In: *Introduction to Analytical Electron Microscopy* (eds. J.J. Hren, J.I. Goldstein and D.C. Joy). Plenum Press, New York, pp. 387–422.

Steeds, J.W. (1984) Electron crystallography. In: *Quantitative Electron Microscopy*. (eds J.N. Chapman and A.J. Craven). Scottish Universities Summer School in Physics, Edinburgh, pp. 49–96.

Steeds, J.W and the Bristol group (1984) *Convergent Beam Electron Diffraction of Alloy Phases* (ed. J. Mansfield). Hilger, Bristol.

Tanaka, M. and Terauchi, T. (1985) *Convergent-beam Electron Diffraction.* JEOL Ltd, Tokyo.

Tanaka, M., Saito, R., Ueno, X. and Harada, Y. (1980) Large-angle convergent-beam electron diffraction. *J. Electron Microsc.* **29**: 408–412.

Thomas, G. and Goringe, M.J. (1979) *Transmission Electron Microscopy of Materials.* John Wiley, New York.

Unwin, P.N.T. and Henderson, R. (1975) Molecular structure determination by electron microscopy of unstained crystalline specimens. *J. Mol. Biol.* **94**: 425–440.

Villars, P. and Calvert, L.D. (1985) *Pearson's Handbook of Crystallographic Data for Intermetallic Phases.* ASM, Metals Park, Ohio.

Wells, A.F. (1984) *Structural Inorganic Chemistry*, 6th Edn. Oxford University Press, Oxford.

Williams, D.B. and Carter, C.B. (1996) *Transmission Electron Microscopy. A Textbook for Materials Science.* Plenum Press, New York.

Willis, B.T.M. (1958) An optical method of studying the diffraction from imperfect crystals. III Layer structures with stacking faults. *Proc. Roy. Soc.* A **248**: 183–198.

Index

Abbe criterion, 2
Aberration, spherical, 7, 9–10, 15–16, 65, 107
  constant, 10
  error in diffraction pattern from,
    9–10
Alignments of instrument, 22–23
Allowed reflections; *see* Systematic absences
Amorphous materials; *see* Diffraction, pattern
  from amorphous materials
Amplitude of diffracted beam; *see* Structure
  factor; Wave, amplitude of
Angle
  Bragg; *see* Bragg angle
  convergence; *see* Convergence angle
  scattering, 10, 81
Antiphase domains, periodic, diffraction from,
  121, 126–127
Aperture
  condenser 1, 4; *see also* Spot size
  condenser 2, 4, 11–12, 15–16, 21, 22,
    98, 106–107; *see also* Convergence angle
    centring of, 16, 21
  diffraction, 6–9, 11, 19, 75, 106–107
  objective, 4–5, 7, 19–22, 65–66
  selected-area; *see* Aperture,
    diffraction
  virtual, 6
Astigmatism, 65
  in the condenser lens, 16, 21–22
  in the diffraction lens, 20, 22–23, 49, 94
    correction of, 22–23
Atomic scattering amplitude, 81–82

Beam
  coherence, 20
  convergence; *see* Convergent beam
    electron diffraction
  damage, 11, 17–18, 37
    displacement, 18
    heating, 17
    ionisation, 17
    knock-on, 18
  diffracted, 1–3, 9–10, 28–31, 56–57,
    71; *see also* Bragg reflection
    amplitude of; *see* Structure factor
    intensity of, 57, 82, 88–89, 114
  incident, 11, 30, 70
  two-beam conditions; *see* Two-beam,
    conditions

  spreading, 16
  stop, 20, 65
  undeviated, 9, 21, 29–30, 34, 52, 55, 77, 98,
    100, 106
Beam-sensitive specimens; *see* Beam damage
Black cross, 105
Bloch waves, 69
Bragg
  angle, 29, 39, 55–56, 65–66, 69–70, 73, 105,
    106
  condition, deviation from; *see*
    Deviation parameter
  Law, 28–31, 56
  reflection, 28, 56
Bravais lattice; *see* Lattice, Bravais
Bright-field (BF)
  disc, 98, 100
  image, 65
  symmetry; *see* Symmetry, in CBED
Buxtons tables, 100–104

Camera constant, 31, 42, 48–50, 93
  calibration of, 49–50
Camera length, 7–8, 17, 20, 22, 23, 30, 93, 102
Coherence; *see* Beam, coherence
Commensurate modulations, 128; *see also*
  Satellites
Composition plane of twin, 62–64
Condenser lens; *see* Lens, condenser
Conjugate plane, 1
Constructive interference; *see* Interference of
  waves
Convergence angle, 11–12, 15, 21, 93, 99
Convergent-beam electron diffraction, 11–16,
  92–108
  patterns, symmetry of, 96–104
  spatial resolution of, 14–16
Cooling the specimen, 11, 17, 20, 94, 131
Cross section, capture, 17
Crystal class; *see* Point group, crystallographic
Crystal system, 133–134
  determination of, 91, 94
  lattice; *see* Lattice, crystal

Damage; *see* Beam damage
Dark-field
  image, 65–66, 113, 117–119
  reflections, internal symmetry of, 102
  tilt controls, 65, 98

167

CPSIA information can be obtained at www.ICGtesting.com
Printed in the USA
LVOW08s2251300615

444450LV00012B/78/P